Governments and Geographic Information

Professor Ian Masser, Department of Urban Planning and Management, International Institute for Aerospace, Survey and the Earth Sciences (ITC), 7500 AA Enschede, The Netherlands, and University of Sheffield, Department of Town and Regional Planning, University of Sheffield, Western Bank, Sheffield, S10 2TN.

Governments and Geographic Information

IAN MASSER
ITC, Enschede
and
University of Sheffield

UK Taylor & Francis Ltd, One Gunpowder Square, London EC4A 3DE
USA Taylor & Francis Inc., 1900 Frost Road, Suite 101, Bristol, PA 19007

British Library Cataloguing in Publication Data

A catalogue record for this book is available from the British Library

ISBN 07484 0789 8 (cased)
ISBN 07484 0706 5 (paperback)

Library of Congress Cataloging in Publication data are available

Cover design by Hybert Design & Type, Waltham St Lawrence, Berkshire.
Typeset in Times 10/12pt by Graphicraft Typesetters Ltd, Hong Kong.
Printed in Great Britain by T. J. International Ltd. Padstow, UK.

Contents

List of figures and tables

FIGURES

COLOUR PLATES

TABLES

Preface

This book is the product of my continuing interest in the processes and consequences of the diffusion of geographic information technologies which was originally triggered off by the publication of the Chorley Report a decade ago. It began with a series of studies evaluating the impact of geographic information systems on British local government, culminating in the publication of a co-authored book with Heather Campbell on GIS and organisations in 1995. The findings of these studies highlighted the rapid rate at which diffusion was taking place throughout the local government sector. At the same time they also demonstrated that the adoption of a new technology and its effective utilisation were two very different things. As a result there is a considerable gap between the expectations expressed in much of the literature and the realities of GIS implementation in a complex organisational environment.

These studies aroused a great deal of interest not only in Britain but also in other European countries and in North America. This stimulated a more ambitious project on the diffusion of geographic information systems in local government in Europe which was undertaken under the auspices of the European Science Foundation's GISDATA scientific programme. This project involved a series of case studies of GIS diffusion in nine European countries as well as an exploration of the broader organisational theoretical perspectives underlying this research and their potential impacts on society. The findings of this project were published in a co-edited book with Heather Campbell and Max Craglia entitled *GIS Diffusion: The Adoption and Use of Geographic Information Systems in Local Government in Europe* in 1996. They showed that, while there were many common features between the experiences of these countries with respect to the recent history of diffusion, there were also some important differences between them. One of the most striking differences concerned the links between digital data availability and the level of GIS diffusion. The findings of the case studies suggested that the question of digital data availability was not simply a matter of the information rich verses the information poor. It was much more a question of central and local government attitudes towards the management of information. Countries with relatively low levels of digital data availability and GIS diffusion also tended to be countries where there had been a fragmentation of data sources in the absence of central or local government coordination. Conversely, countries where government had created a framework in terms of responsibilities, resources and standards for the collection and management of geographic information also tended to be those with relatively high levels of digital data availability and GIS diffusion.

Given these findings, I decided to explore in greater depth the role that governments play with respect to geographic information. The need for some form of national geographic information strategy has been a recurring theme in geographic information circles ever since the publication of the Chorley Report and a number of countries have already taken steps to formulate elements of such a strategy. The most publicised of these initiatives is the US National Spatial Data Infrastructure which was established in 1994 as a result of an Executive Order signed by President Clinton himself. The obvious thing to do then was to visit some of these countries to try to put together a reasonably consistent picture of recent developments.

I was able to turn this intention into reality as a result of a number of events. First, I was awarded an ESRC Senior Research Fellowship (H52427501895) to carry out research on building the European geographic information resource base. This relieved me from my teaching and administrative duties at Sheffield for the 1995/6 academic year as well as providing me with the resources that I required to carry out the fieldwork in Britain and the Netherlands and at the European Commission. Second, between the submission of my ESRC fellowship application in November 1994 and the commencement of the fellowship in September 1995 I was involved in several rounds of the consultations that took place on the proposals put forward by DGXIII for a European Geographic Information Infrastructure. This proved invaluable in developing my own thinking on the nature of the key issues involved. Third, I was invited by the Australasian Urban and Regional Information Systems Association to deliver a keynote paper at their annual conference in Melbourne in November 1995. This invitation not only gave me an opportunity to visit Australia but also came with the condition that I visit the AURISA chapters in Adelaide and Canberra to meet key figures in the geographic information field at the regional as well as the national levels. The visit proved very enjoyable but was also a very effective way of carrying out an Australian case study. Finally, I was fortunate to be involved in a very productive collaboration with Bob Barr from the University of Manchester in early 1996 which enabled me to develop my ideas on the broader theoretical issues involved in the relationship between governments and geographic information. A great deal of the material that is presented in Chapter 2 of this book is derived from this collaboration and first saw the light of day in a joint paper entitled ‘Geographic information: a resource, a commodity, an asset or an infrastructure?’, which was presented to the fourth annual GISRUK conference in Canterbury in April 1996.

The rest, as they say, is history. However, it must be emphasised that the existence of this book is very heavily dependent on the encouragement and assistance that I have had from many people. I am particularly indebted to Mike Blakemore, Denise Lievesley and David Rhind for their support throughout the project as well as for their helpful and constructive comments on various drafts of the text. Given the nature of the research and the central role played by the four case studies of Britain, the Netherlands, Australia and the United States, I have been particularly fortunate in the assistance that I have received from key individuals in these countries. I would especially like to thank Mike Blakemore, Alan Oliver, David Rhind and Peter Smith for their valuable comments on earlier drafts of the British case study. Henk Scholten played an important role in setting up the interviews for the Dutch case study. He also made a number of very constructive comments on early drafts of the text, as did Jaap Berends, Hans van der Linde, Dick Meuldijk and Paul van der Molen. Graham Baker, Allan Barnes and Ian Williamson performed a similar role in connection with the Australian case study. Eric Anderson helped to set up the interviews for the US case study. He also reviewed several drafts of the chapter for the United States, as did Joel Morrison, Harlan Onsrud, Karen Siderelis and Nancy Tosta. Needless to say, I must point out that none of these individuals can be held responsible for any of the interpretation that I have given to events in the text. This must remain entirely my responsibility.

I must also thank Max Craglia who helped to keep things on an even keel in Sheffield during my frequent absences, and express my gratitude to Dale Shaw who carried out the thankless task of seemingly endless revisions to the manuscript in a thoroughly professional way. Last but not least, I must thank my wife, Suzy, for her patience and forbearance during the preparation of this book. I dedicate it to her with love and affection.

IAN MASSER
1997

CHAPTER ONE

The Emergence of National Geographic Information Strategies

INTRODUCTION

It is now more than 10 years since the Committee of Enquiry on Handling Geographic Information, chaired by Lord Chorley, described geographic information systems (GIS) as 'the biggest step forward in the handling of geographic information since the invention of the map' (Department of Environment, 1987, para. 1.7). Since then the number of GIS applications throughout the world has increased exponentially and now extends from local banking and business services to large-scale utilities management and from urban planning to global environmental modelling. As a result, GIS has become a multimillion-dollar industry whose development impacts on many different aspects of economic and social life.

The adoption and use of GIS technology has placed new demands on existing sources of geographic information. It has transformed conventional mapping and cartography out of all recognition. In place of the traditional paper map the new technology makes it possible to integrate digital geographic information from a variety of different sources to specifications designed by users rather than producers. In the process the same databases are often drawn on again and again in a wide range of different applications.

However, as the Chorley Committee pointed out, the availability of the technology itself is a necessary but not a sufficient condition for its effective utilisation. To facilitate the take-up of GIS a number of institutional barriers must be overcome. Of particular importance in this respect is the need to take steps to promote the availability of digital data in forms that facilitate its use in a wide range of applications.

Issues such as these raise important questions regarding the future role that governments will play with respect to geographic information, as it must be recognised that, with the exception of one-off research activities, most operational applications of GIS are in some measure dependent upon the availability of data collected by government agencies. As Rhind (1995a, p. 101) has pointed out:

> Government is almost everywhere the primary sponsor at present of the national geodetic framework and the source of 'framework' data provided by the topography (used in the widest sense). It is also typically the source of geological, soils, meteorological, pollution, demographic, land ownership, taxation, employment and unemployment and many other national datasets, including statistical time series. Virtually all of this data were originally collected for the purposes of the national state or subsets of it.

Consequently, it must be recognised that the steps taken by these government agencies to make their data available in forms that facilitate its integration with other data sources are likely to have a profound effect on the extent to which GIS technology is utilised in different countries. This view is borne out by the findings of a recent comparative analysis of the adoption and utilisation of GIS in local government in nine European countries which suggest that the level of GIS diffusion in these countries is closely linked to the availability of digital data (Masser, Campbell and Craglia, 1996). The dissemination of geographic information is also influenced by a wide range of indirect factors over which governments are able to exert varying degrees of control. Rhind (1996a, p. 8) calls this 'a cocktail of laws, policies, conventions and precedents which determine the availability and price of spatial data'.

Given these circumstances, many governments throughout the world are starting to think more strategically about geographic information. The most highly publicised initiative is the programme to establish a National Spatial Data Infrastructure in the United States to coordinate geographic data acquisition and access which was launched by an Executive Order from the White House signed by President Clinton himself on 11 April 1994 (Executive Office of the President, 1994). This outlines the measures that need to be carried out and sets a first-stage timetable for their implementation. The rationale underlying such measures is summarised in the following terms in the preamble to the Executive Order: 'geographic information is critical to promote economic development, improve our stewardship of national resources and protect the environment'.

Similar but less publicised initiatives are being launched in many other countries. For example, the Korean government (Ministry of Construction and Technology, 1995) recently announced that it is allocating \$360 million over five years for the development of a National Geographic Information System on the grounds that this 'is recognised as one of the most fundamental infrastructures required in promoting national competitiveness and productivity'. In Japan a special association consisting of representatives from major industrial interests such as Mitsubishi, Sumitomo and Tokyo Marine Insurance has been set up to promote a national spatial data infrastructure, and the government is taking steps to improve its digital topographic database at a cost that may amount to around \$1.5 billion over the next five years (Yamaura, 1996).

Some countries have already set up coordinating bodies to formulate and implement national geographic information strategies. These include the Australian Land Information Council which was established in January 1986 by agreement between the Australian prime minister and the heads of state governments to coordinate the collection and transfer of land-related information between the different levels of government and to promote the use of that information in decision-making. In November 1991 New Zealand became a full member of the council which was renamed the Australia New Zealand Land Information Council (ANZLIC). ANZLIC's vision is set out in its strategic plan for the period 1994–7 as follows: 'Australia and New Zealand will have the land and geographic data infrastructure needed to support their economic growth, and their social and environmental interests, backed by national standards, guidelines, and policies on community access to that data' (ANZLIC, 1994, p. 5).

The Dutch Council for Real Estate Information (Ravi) was set up in 1984. Its initial role was to advise the minister for housing, spatial planning and the environment on matters relating to the operation of the Cadastre. In 1990 the Council was asked to review its position in the light of the increasing computerisation of real estate and geographic information services and the growing autonomy of the Cadastre. It produced a master plan, setting out its future role as a National Council for Geographic Information. In 1995

Ravi published a discussion document outlining its plans for a National Geographic Information Infrastructure to create economic opportunities, facilitate spatial planning and investment, and promote sustainable development (Ravi, 1995c).

Matters relating to geographic information also featured on the agenda of the G7 Ministerial Conference in Brussels in February 1995 and have stimulated discussions within the European Union regarding the need for a European Geographic Information Infrastructure. As in most national initiatives, economic competitiveness and the need for more effective planning and investment feature prominently in the list of expected benefits to be derived by the implementation from a strategy of this kind:

- efficiencies of scale in a unified market;
- reduced problems for trans-border and pan-European projects;
- new business opportunities for the European geographic information industry;
- ability to design technical solutions for future growth;
- increasing use of European skills and improved market position in geographic information;
- improved capability for European-wide planning and decision making.

(DGXIII, 1996, p. 14)

Objectives of the Book

Despite the interest that has been aroused by such developments, the literature that is available on them in hard-copy form or via the World Wide Web at the present time is largely descriptive in character and partial in nature. Typically it outlines some or all of the measures proposed in one particular country without reference either to the administrative circumstances governing geographic information provision in that country or the broader institutional contexts within which the measures have been developed. This makes it very difficult to assess these national experiences in anything like a systematic way and also gives rise to major misunderstandings when the experience of one country is evaluated by someone from another country where there is a very different situation regarding the provision of geographic information.

For this reason it can be argued that the time is now ripe for a more comprehensive analysis of these developments than has hitherto been the case. With these considerations in mind, this book considers the nature of the relationship between governments and geographic information from a number of different conceptual positions and also with reference to the experiences of four different governments in terms of the development of national geographic information strategies. The discussion is divided into three main parts. The first of these considers what is special about geographic information and evaluates the notion of geographic information from four different standpoints: as a resource, a commodity, an asset and an infrastructure (Chapter 2). The second part presents the findings from four case studies of national geographic information strategies in Britain, the Netherlands, Australia and the United States (Chapters 3–6), while the final section evaluates these experiences with a view to identifying what general lessons can be learnt from them (Chapter 7).

The main objective of this book is to examine the role that governments can (or could) play in shaping the development of national geographic information strategies directly through a variety of policy initiatives and also indirectly because of the extent to which they create the broader institutional context within which these are developed and implemented. Consequently the main emphasis of the discussion is on policy rather than technical considerations. For this reason it contains relatively few details regarding either the specification of the systems developed by the main providers of geographic information or the standards used to facilitate data transfer and exchange. It should also be noted that the focus of the analysis is international and comparative in nature. As a result the turf wars between different government departments and the power struggles surrounding geographic information in particular countries are touched on to a much lesser extent than would be the case in a single-country case study of national geographic information strategy.

Prior to the detailed discussion it is necessary to clarify what is meant by the terms 'geographic information' and 'national geographic information strategies', and also to explain the reasons behind the choice of case studies and the basic analytical framework used for these studies.

SOME DEFINITIONS

In this book the term 'geographic information' refers to 'information that identifies the geographic location and characteristics of natural or constructed features and boundaries on the earth' (Executive Office of the President, 1994, Section 1b). It carries essentially the same meaning as the terms 'spatial data', 'geospatial information' and 'geospatial data' which are also widely used in the literature.

The term 'national geographic information strategies' is used in preference to 'national spatial data infrastructure' or 'national geographic information infrastructure' to describe 'the technology, policies, standards and human resources necessary to acquire, process, store, distribute, and improve utilisation of [geographic information]' (Executive Office of the President, 1994, Section 1a). In this respect the term 'national' refers essentially to government initiatives. There are also a few cases of supranational initiatives of this kind. A good example of this is the policy framework for geographic information that is being considered within the European Commission (DGXIII, 1996, p. 11) which is seen as 'a set of agreed rules, standards, procedures, guidelines and incentives for creating, collecting, exchanging and using geographic information'.

The term 'strategy' is used in preference to the term 'infrastructure' for several reasons. First, as will be explained in Chapter 2, while the concept of infrastructure invites some fruitful comparisons with other types of infrastructure such as transport networks or schools which are regarded as fundamental to the effective functioning of society, it is felt that the term itself does not encompass the combination of technologies, policies, human resources and knowledge that is implied by either of the definitions quoted above. Second, it should also be noted that national geographic infrastructures already exist in every country to a greater or lesser degree. What is also implied by the term 'national geographic information strategy', then, is the need for government to take active steps to improve these infrastructures to make them more responsive to the new demands for geographic information created by the advent of GIS technology.

Most statements regarding the need for such strategies emphasise one or more of the following objectives: to promote economic competitiveness, to improve the quality of decision-making and to provide better stewardship of national resources and the environment (see, for example, Executive Office of the President, 1994; ANZLIC, 1994, p. 5; Ravi, 1995c, pp. 4–5; DGXIII, 1996, p. 12). However, it should also be borne in mind that a successful national geographic information strategy may have negative as well as positive consequences for society as a whole. Some of these consequences have been explored by Wegener and Masser (1996) who argue that the view that GIS in Europe will be universal in extent and largely benign in operation in 20 years' time is only one of a number of possible brave new GIS worlds scenarios. Current trends could also lead to a market-driven scenario in which geographic information is bought and sold like any other commodity. In this scenario governments could lose their control over information and find themselves at the mercy of multinational information conglomerates. As a result there may be marked gaps between the information rich and the information poor, and between providers and consumers of information. Alternatively, Wegener and Masser argue that current trends may result in a scenario in which the European corporate state makes extensive use of information to defend itself and its citizens against crime and subversive activities. In this Big Brother scenario there is a marked gap between those who benefit from the greater security provided by these surveillance activities and the illegal immigrants and the tax evaders whose lives are under constant scrutiny because of the threat they pose to this security.

THE CHOICE OF CASE STUDIES

The choice of case studies was partly dictated by practical considerations and partly by prior knowledge of the countries concerned with respect to the development of national geographic information strategies. From the outset it was recognised that language was likely to be an important factor in the choice of case studies, given that most of the materials that would be needed for them would only be available in the language of the country itself. Consequently the author was restricted by necessity to English- or Dutch-speaking countries at the expense of some potentially interesting case studies such as Portugal's Sistema Nacional de Informacio Geografica.

Australia, Britain, the Netherlands and the United States were also selected as case studies because they can be regarded as among the leaders in the field. Australia has a tradition of Commonwealth land information management coordination dating back 10 years. Britain recently completed its national digital topographic database. Like Australia, the Netherlands has a tradition of geographic information coordination activities while the high profile of the National Spatial Data Infrastructure in the United States has attracted a great deal of attention from all parts of the world.

The four countries selected represent a wide range of contexts for the development of national geographic information strategies. Britain and the Netherlands are small, densely populated countries with a strong mapping tradition dating back several centuries. In contrast, the United States and Australia are much bigger countries containing large parts that have been settled only relatively recently. Unlike Britain and the Netherlands, both these countries have federal systems of government whereby many of the responsibilities carried out elsewhere by central government are decentralised to state or even local governments. As might be expected, the duties of the government agencies, which are the main providers of geographic information in these countries, vary considerably, as does the legal and institutional context within which national geographic information strategies have come into being.

The main objective of the case studies is to consider the steps that are currently being taken to implement national geographic information strategies in their national contexts. The purpose of this evaluation is neither to create a league table of national performance nor to produce cookbook recipes for other countries to follow. It is essentially to identify what general lessons for other countries might be learnt from the four national experiences.

How far these lessons may be appropriate for transfer to other countries must remain a matter of conjecture which can only be resolved by reference to their particular institutional and political circumstances.

ANALYTICAL FRAMEWORK

To facilitate comparative analysis a common analytical framework is used for the presentation of the findings of the four case studies. This consists of the five parts set out in Table 1.1. The first of these provides a brief introduction to the geographical and historical context of each country. The second describes the activities of the main providers of geographic information in the public sector as a whole. It was recognised at the outset that a large number of public agencies are involved in the provision of various kinds of geographic information. These include cadastral and mapping agencies, national statistical organisations, a wide range of environmental bodies including the agencies responsible for geological and hydrographical mapping, various local government organisations, as well as agencies involved in defence activities and emergency management. For practical reasons, given the number of agencies involved, a decision was made to limit the analysis to the agencies that are responsible for land titles and cadastral information, large- and small-scale mapping activities and the provision of socio-economic statistics for geographic areas.

The third part of the framework considers the institutional context within which national geographic information strategies are being developed in each of the four countries. In practice, this involves two main issues: overall government polices (both explicit and implicit) with respect to the collection and dissemination of geographic information, and matters relating to the legal protection of geographic information in each country with particular reference to intellectual property rights and privacy.

The fourth part examines three key elements of national geographic information strategy: the organisational arrangements that have been made for coordination, the creation of core data sets, and the establishment of metadata facilities at the national level. Finally, the last part contains a critical evaluation of the progress that has been achieved in each country with respect to the development of a national geographic information strategy.

Table 1.1 Analytical framework used for the case studies

1. GEOGRAPHIC AND HISTORIC CONTEXT
2. MAIN DATA PROVIDERS
   Land titles
   Large- and small-scale mapping
   Socio-economic statistics
3. INSTITUTIONAL CONTEXT
   Dissemination
   Legal protection
4. NATIONAL GEOGRAPHIC INFORMATION STRATEGY ELEMENTS
   Coordination
   Core data
   Metadata
5. EVALUATION

CONCLUSION

The advent of modern GIS technology has transformed spatial data-handling capabilities and made it necessary for governments to rethink their roles with respect to the supply and availability of geographic information. As a result, many governments throughout the world are starting to think more strategically about geographic information and a number of countries have already set up coordinating bodies for this purpose. This book considers the nature of the relationships between government and geographic information both from a number of conceptual standpoints and also with reference to the experience of four different governments in terms of the development of national geographic information strategies. The discussion that follows is divided into four main parts. Chapter 2 considers what is special about geographic information and evaluates it from four different standpoints. Chapters 3 to 6 present the findings of the four case studies, while Chapter 7 evaluates them with the objective of identifying what lessons might be learnt from these experiences.

CHAPTER TWO

What is Geographic Information and why is it so important?

INTRODUCTION

This chapter considers the nature of geographic information from a number of different standpoints, drawing in the process on the general literature surrounding the debate on information as a whole as well as on literature from the geographic information field. The first part considers what is special about information and geographic information both in the overall context of the emergence of a global information economy and with particular reference to the significance of geographic location as a means of integrating databases held by a large number of public and private agencies.

The remainder of the chapter considers the role of government with respect to geographic information from four different conceptual positions: as a resource, a commodity, an asset and an infrastructure. In the process three key elements of national geographic information strategies will be identified: the need for overall coordination, the significance of metadata services, and the strategic importance of a relatively limited number of core databases. The final section of the chapter summarises the main findings that emerge from this analysis.

THE SIGNIFICANCE OF INFORMATION AND GEOGRAPHIC INFORMATION

Dictionary definitions of information are not entirely helpful because there is a degree of circularity in the relationships between knowledge, information and data.

In a geographical context it is perhaps easiest to think of a hierarchy. At the bottom of the hierarchy comes data. Data can be thought of as the symbolic representation of some set of observations. Data comprise the raw characters and digits that occupy computer storage media. Data alone have little value, but incur significant storage costs and some transmission costs. At the next level in the hierarchy comes information. Information implies both data and context. The context may be the definitions of the variables that the data represent, their reliability and timeliness. These additional characteristics of data are usually termed 'metadata'. In simple terms, for most information, data + metadata = information. Knowledge in turn implies understanding. For information to become knowledge, an element of understanding of the significance and the behaviour of that information is required. In the academic world, data and information alone are considered to be of little value except as a step on the path to knowledge. However, in everyday life information alone is usually sufficient. It is enough to know that a postal address, or a house number and a postcode, refers to an addressable property. No deeper significance needs to be sought.

As raw data are of no value, and true knowledge is seldom attained, much of the argument revolves around information. It is also useful to consider the concept of information in the context of the information economy. Goddard (1989, pp. xvi–xvii) argues that this contains four propositions:

- information is coming to occupy centre stage as a key strategic resource on which the production and delivery of goods and services in all sectors of the world economy is dependent . . . ;
- this economic transformation is being underpinned by technological transformation in the way in which information can be processed and distributed . . . ;
- the widespread use of information and communications technologies is facilitating the growth of the so-called tradeable information sector in the economy; and
- the growing informatisation of the economy is making possible the global integration of national and regional economies.

This provides a useful starting point for the general discussion in that it emphasises the linkages between information as such and the communications technologies that have recently come into being. It is also valuable in that it draws attention to some of the implications of these developments for economic restructuring and globalisation. General support for these propositions can be found in the work of Harlan Cleveland (1985, p. 185): 'information (organised data, the raw material for specialist knowledge and generalist wisdom) is now our most important and pervasive resource'.

Geographic information must be seen as a special case of information as a whole. In the previous chapter it was defined as 'information that identifies the geographic location and characteristics of natural features and boundaries on the earth' (Executive Office of the President, 1994, Section 1b). This definition makes the distinction between two types of geographic information: location and attribute. Location is clearly essential in order to make information geographic. However, locational information without attribute information is of little inherent interest.

The economic significance of geographic information lies in the general referencing framework that it provides for integrating large numbers of different data sets from many application fields in both the public and private sectors. For this reason items of geographic information such as topographic maps, standardised geographic coordinate references such as the National Grid, standardised geographic referencing systems such as street addresses, and standardised areas such as administrative subdivisions and postcode sectors are particularly valuable as they make it possible to link different data sets and thereby to gain additional knowledge from them. An early example of adding value in this way was the work of the London doctor John Snow, who demonstrated in 1854 that there was a strong association between the home locations of cholera victims and the water pump in Broad Street by combining two data sets, deaths and water pumps, on a single map (Gilbert, 1958).

Although the value of linking such data has long been recognised, it was not until recently that computer technologies came into being that are capable of manipulating large quantities of geographic data in digital form. With the arrival of geographic information systems handling technology during the 1980s, the potential for linking geographic data sets increased dramatically and the modern geographic information economy came into being.

> Unlike maps, strings of geographical spatially referenced data can be aggregated, transformed, and shared. Spatial data can now be more easily isolated and abstracted from the particular application in which it was developed and channelled into other settings and other GIS where it can be reused, enhanced, and routed to other potential user communities.
> (National Research Council, 1993, p. 8)

It has been estimated that between 60 and 80 percent of all data held by British government agencies can be classified as geographical in this respect (Nansen, Smith and Davey, 1996).

It is important to note that those elements of geographic information which are of special significance are also those that require a considerable measure of consistency to be used effectively. However, in practice, map projections may vary from one country to another and in different applications. Similarly the administrative areas used for local government are not necessarily the same as those used for other purposes. Furthermore, subdivisions such as these vary over time as modifications are made for administrative purposes.

Consequently it must be recognised that there is a need for some measure of standardisation if the potential for linking data sets is to be fully exploited. This means that units such as postcodes, which were originally intended to facilitate the operations of the postal services, acquire new significance for geographic information handling.

These features of geographic information make it not only a special case of information in general but also of the new information economy that is transforming society.

GEOGRAPHIC INFORMATION AS A RESOURCE

Both Goddard and Cleveland refer to information as a resource which has many features in common with other economic resources such as land, labour and capital. However, Cleveland (1985) also argues that information possesses a number of characteristics that make it inherently very different from these traditional resources. As a result, the laws and practices that have emerged to control the exchange of these resources may not work as well (or even at all) when it comes to the control of information:

Cleveland identifies six unique and sometimes paradoxical qualities of information which make it unlike other economic resources:

> 1. Information is expandable, it increases with use.
> 2. Information is compressible, able to be summarised, integrated, etc.
> 3. Information can substitute for other resources, e.g. replacing physical facilities.
> 4. Information is transportable virtually instantaneously.
> 5. Information is diffusive, tending to leak from the straightjacket [sic] of secrecy and control and the more it leaks, the more there is.
> 6. Information is shareable, not exchangeable, it can be given away and retained at the same time.
> (Cleveland, cited in Eaton and Bawden, 1991, p. 161)

Given the significance attached to these six qualities in much of the literature on the economics of information, it is useful to consider them in some detail with particular reference to their applicability to geographic information.

Information is Expandable

Mason, Mason and Culnan (1994, p. 42) point out that information tends to expand with its use as new relationships and possibilities are realised. Cleveland (1982, p. 7) also draws attention to the synergetic qualities of information: 'the more we have, the more we use and the more useful it becomes'.

There are close parallels between these views and those of leading writers on geographic information. For example, David Rhind (1992, p. 16) concludes that 'all GIS experience thus far strongly suggests that the ultimate value is heavily dependent on the association of one data set with one or more others; thus in the EEC's CORINE (and in perhaps every environmental) project, the bulk of the success and value came from linking data sets together'. However, Rhind also emphasises that this depends very much on the unique properties of the geographic information referencing system that is referred to above. 'Almost by definition, the spatial framework provided by topographic data is embedded in other data sets (or these are plotted in relation to it, or both); without this data linkage, almost no other geographical data could be analysed spatially or displayed' (p. 16).

Paradoxically, geographic data often expand, while being degraded. Census data, for example, is anonymised by aggregation, but the process of aggregation and tabulation actually increases the size of the data set and this increase is potentially almost infinite. A small number of census questions posed to individuals and households yields over 8,000 counts when aggregated up to the level of the enumeration district and tabulated. The tables that are produced are only a subset of all those that could be produced. Thus the process of geographical analysis and re-aggregation itself has the potential to expand information.

Similarly, the editors of a major work on sharing geographic information (Onsrud and Rushton, 1995, p. xiv) point out that 'sharing of geographic information is important because the more it is shared, the more it is used, and the greater becomes society's ability to evaluate and address the wide range of problems to which information may be applied'.

Information is Compressible

Another unique feature of information is the extent to which complex data sets can be summarised. Cleveland (1982, p. 8) points out: 'We can store many complex cases in a theorem, squeeze insights from masses of data into a single formula [and] capture lessons learned from much practical experience in a manual of procedure.' The enormous flexibility of information and the opportunities it opens up for its repackaging in different ways to meet the demands of particular groups is clearly evident in current geographic information practice. A good example is the field of geodemographics which makes extensive use of a wide range of lifestyle classifications derived from a mass of small-area census statistics. This reverses the process of expansion described above.

In addition to the compression of attribute information, referencing information can also be compressed. For example, given a suitable look-up table, a personal name and a post code in Britain almost always uniquely identifies a postal address delivery point. It is important to distinguish between application-specific means of compression, where geographic characteristics determine the form of compression used, and generic compression that can be used on any digital data.

Information is Substitutable

In essence this property of information means that it can replace labour, capital or physical materials in most economic processes. According to Mason, Mason and Culnan (1994, p. 44), substitutability is one of the main reasons for information's power and it can be used to harm as well as to improve the human condition.

Once again there are many examples of substitutability in the geographic information field. A typical example is the requirement placed on British local authorities by the 1991 New Roads and Streetworks Act to maintain a computerised street and roadworks register. The geographic information contained in such a register makes it possible to locate underground facilities more precisely than before, thereby saving the time needed by workers to carry out essential maintenance and repairs as well as reducing the number of disruptions caused by the works.

Information is Transportable

As a result of the digital networks that have been created by modern technology, large quantities of information can be transferred from one place to another in the world almost instantaneously. This means that information users can create their own virtual databases by accessing the data that they need from other databases as and when they are needed. A good example of a virtual database for geographic information is the National Land Information Service pilot project for Bristol in Britain, which operates on the Land Registry's mainframe and provides on-line access to live databases held by Ordnance Survey at Southampton, the Valuation Office at Worthing, the Land Registry at Plymouth and Bristol City Council at Bristol (Manthorpe, 1995).

Information is Diffusive and Information is Shareable

These two properties have a number of common features and are best dealt with at the same time. The idea that information is diffusive reflects the intangible qualities of information which distinguish it from material resources. The idea that information is shareable draws attention to the extent to which information can both be given away and retained and does not wear out as a result of being used. Both these ideas have been given new significance by recent technological developments which make it possible to copy information at near zero cost. Both raise fundamental problems about the ownership of digital information. For this reason Anne Wells Branscomb (1995, p. 17) argues that traditional ways used to control the flow of information such as copyright need rethinking in the context of digital information. This is because the roots of the paradigm of copyright are embedded in the printing press and the notion that there is an artefact that can be copied. 'In the computer environment it is access to organised information which is valuable and everything is copied. It is impossible to use a computer programme without copying it into the memory of the computer'. Because of its diffusability and shareability, information has many features of a public good in economics: i.e. its benefits can be shared by many people without loss to any individual and it is not easy to exclude people from these benefits (see, for example, Coopers and Lybrand 1996, pp. 11–13; and Perritt, 1996, p. 419).

Similar views have also been expressed in the debates regarding geographic information. For example, Rhind (1992) notes that geographic information does not wear out with use although its value may diminish over time due to obsolescence, and it can be copied at near zero cost although this may not be the case in instances where currency is important. Rhind also identifies two conflicting tendencies on the international scene which reflect these properties: the increasing commercialisation of geographic information supply and the free exchange of data among scientists working on global problems.

Onsrud (1995, p. 293) considers this situation from the standpoint of power:

> Because geographic information has potential value to those with effective access to it, this realisation gives rise to the desire to exercise ownership rights over this information . . . this desire to control information is in direct tension with recent technological realities that make the copying, dissemination and sharing of information very inexpensive . . As technology improves, this tension will only increase.

The above discussion highlights some of the parallels that exist between the issues that are being debated in the geographic information field and those identified by Harlan Cleveland as unique properties of information as a whole. Because of these properties, information cannot be treated as a resource in the same way as land, labour or capital. Nevertheless, as Eaton and Bawden (1991, p. 165) point out, information must still be regarded as a resource in the sense that it is vital to organisations because of its importance to the individuals within them. The task for managers is therefore to exploit the potential of information as a resource while taking account of its singular qualities.

GEOGRAPHIC INFORMATION AS A COMMODITY

The concept of the information economy assumes that information can be bought and sold like any other commodity. Yet, as was the case with the notion of information as a resource, its unique features make it significantly different from other commodities. This is particularly apparent in the public good dimension that information possesses because of its shareability and its diffusiveness. Unlike other commodities, information remains in the hands of the seller even after it has been sold to a buyer and its inherent leakiness makes exclusive ownership of information problematic. In addition, the compressibility of information makes it difficult to define what constitutes a unit of information and its expandability raises questions about how to gauge its value given that this depends heavily on its context and its use by particular users on particular occasions (Repo, 1989).

In summary, then:

> It is fashionable to speak of information as a commodity, like crude oil or coffee beans. Information differs from oil and coffee, however, in that it cannot be exhausted; over time certain types of information lose their currency and become obsolete, but, equally certain types of information can have multiple life cycles. Information is not depleted by use, and the same information can be used by, and be of value to, an infinite number of consumers.
>
> (Cronin, 1984)

The discussion of information as a commodity is further complicated by the fact that large quantities of information are collected by government agencies in the course of their administrative duties and most of this information is not made available to the public. This is particularly the case with respect to geographic information. In Britain, for example, nearly 600 separate data sets held by government agencies are listed in Ordnance Survey's Spatial Information Enquiry Service (SINES) but many of these are not available to external users.

There are a number of reasons why information collected by government is not made generally available. In the first place, it can be argued that some of this information is obtained on condition that its confidential nature will be respected. In the second place, it can also be argued that personal information acquired for the performance of a specific statutory duty should not be made available for other purposes. Finally, it can be argued that the primary task of government agencies is to carry out their statutory responsibilities and that the dissemination of information collected in the course of these duties may be both a burden on resources and a distraction from their primary responsibilities.

These arguments have been criticised both on the grounds of the needs of the information economy and the need for open government. With respect to the former, Openshaw and Goddard (1987), for example, draw attention to the emerging market for geographic information in the private sector and urge the public sector to make more data holdings available to promote the expansion of this important market. With respect to the latter, Onsrud (1992, p. 6) claims that: 'All other rights in a democratic society extend from our ability to access information. Democracy can't function effectively unless people have ready access to government information to keep government accountable.'

Given these criticisms, there has been much debate about how the costs of collecting, maintaining and disseminating these data sets to the public at large should be paid for. The traditional view is that governments collect information primarily for their own administrative purposes and that any costs they incur in the process are offset by the benefits obtained by the public in the form of a better delivery of public services. Under these circumstances it is argued that, as the public has already paid for the collection of this information through its taxes, it should be made available to the public at no more than the marginal costs of reproduction. It is claimed that such a policy would not only increase the accountability of government departments to the public but also stimulate the growth of the information economy.

The case for and against these arguments has been widely debated over the last few years (see, for example: ALIC, 1990d; Maffini, 1990; Blakemore and Singh, 1992; Onsrud, 1992 and 1995; Rhind 1992 and 1996a). The arguments against public domain databases have been summarised by Rhind (1992, p.17) in five main points:

> First is that only a small number of citizens may benefit from the free availability of data which has been paid for by all and this is not fair. The second argument partially follows on from the first and is that any legal method of reducing taxes through the recouping of [public] expenditure is generally welcomed by citizens. The third is that the packaging, documentation, promotion and dissemination of data invariably costs considerable sums of money . . . thus nothing can be free. The fourth argument in favour of charging is that putting a price on information inevitably leads to more efficient operations and forces consumers to specify exactly what they require. Finally, making data freely available is liable to vitiate cost sharing agreements for its compilation and assembly.

On the other hand, opponents of cost recovery such as Onsrud (1992, p. 6) claim that this runs counter to the principle of public accountability:

> Because of the need to maintain confidence in our public administrators and elected officials and to avoid accusations that they may be holding back records about which citizens have a right to know, we should not restrict access to GIS data sets gathered at tax payer expense simply because that information is commercially valuable to the government.

Onsrud also argues that cost recovery will increase bureaucratic costs and discourage the sharing of geographic information. Furthermore, he claims that cost recovery will result in the creation of government-sanctioned monopolies, and that once these monopolies come into being, there will be little incentive for them to improve the service.

Arguments such as these highlight some of the problems associated with treating information as a commodity, particularly where government agencies are involved. This presents particular problems in the field of geographic information because of the large number of data sets involved and the vital importance of some data sets such as the topographic data collected by national mapping agencies for linking together other data holdings. It should also be noted that national mapping agencies differ from most other government agencies in one significant respect. They do not collect data in order to carry out their administrative duties. Their administrative duties are primarily to maintain the national topographic database (Rhind, 1991).

In summary, then, information has some of the features of a conventional commodity which can be bought or sold. However, it also has a number of unique qualities which must be taken into account in the process. These are closely linked to the previous discussion regarding the notion of information as a resource. However, when it comes to information collected by government agencies, it is necessary to add a number of additional qualifications to the concept of information as a commodity. This includes the recognition that much information is not primarily collected for resale as a commodity and that the costs of much of this information are met through taxes on the public in the interests of good governance.

GEOGRAPHIC INFORMATION AS AN ASSET

The notion of geographic information as a commodity deals with the role of government in general terms with reference to the principle of public accountability while the arguments for and against cost recovery revolve essentially around the extent to which users rather than tax payers should finance the costs incurred by specific government agencies with respect to database creation and maintenance. For this reason the notion of information as an asset introduces an important new dimension regarding the role of government into the discussion. This is the idea that geographic information must be regarded as an asset that needs to be managed effectively in the national interest. In this respect the national interest is usually defined in general terms with reference to the information that is required to meet the needs of those responsible for national defence and emergency services as well as that required for effective public administration and maintaining economic competitiveness. In terms of mapping, three separate components of the national interest have been defined by Ordnance Survey (1996d, p. 5):

- The public interest arising from the mapping of areas which would not otherwise be mapped if the judgement was made solely in terms of revenue generated by sales of that mapping alone. This is particularly crucial in regard to contingencies where there is typically no time to create new mapping: the information must be available 'off the shelf' . . . ;
- The benefits of having consistency of content, currency, style and manner of mapping which is dictated by needs other than those of the local market. Two different categories of this exist: these are where the information is needed for defence purposes and where sizeable external economic benefits occur (such as everyone using the same topographic framework and hence all of the data collected 'fitting together' correctly);

- The inescapable commitment for the creation or maintenance of the underpinning infrastructure of the mapping (notably the geodetic framework) which is widely used by other bodies and by the public and where charging for use is either inappropriate or impossible (such as for use of the National Grid).

The notion of geographic information as an asset also introduces the concept of the custodianship of public information. In the opinion of the Australian Land Information Council (ALIC, 1990a, p. 5), 'the principle of custodianship lies at the core of efficient, effective and economic land information management'. In simple terms,

> All data collected by a state government agency forms part of a state's corporate data resource. Individual agencies involved in the collection and management of such land related data are viewed as custodians of that data. They do not own the data they collect but are custodians of it on behalf of the state. (p. 1)

The Australian Land Information Council (ALIC, 1990a, p. 3) also defines the responsibilities of the custodian:

> These are that the custodian should be responsible for principles and procedures for the accuracy (integrity), currency (timeliness), data storage (i.e. definition and structure) and security of a data item or data collection. In so doing, the custodian must consult with, and take account of the needs of users other than itself.

The concept of custodianship is significant in that it highlights the importance of defining rights of access to database and specifying the responsibilities of custodians for database maintenance. In essence, then, 'the distinction between ownership and access is quite clear. The issue is how to ensure effective access to information of importance to the community for a variety of purposes' (Epstein and McLaughlin, 1990, p. 38).

Both custodianship and the parallel notion of stewardship assume that government information assets are distributed across a wide range of public agencies rather than concentrated in a single body. For this reason, it has been argued that 'the challenge faced by all levels of government is to place data stewardship responsibility as close to the data originator as possible while maintaining an effective national infrastructure' (National Research Council, 1994, p. 14).

This statement highlights the extent to which the notion of geographic information as an asset implies two very different levels of government activity. Those responsible for the creation and maintenance of particular databases have a duty as custodians of these, while at a higher level there is also a need for governments to take steps to ensure that the activities of these custodians reflect the national interest as a whole. This is especially important in the case of geographic information because of the significance attached to the expandability property and the integration of different data sets. For this reason the notion of geographic information as an asset identifies two key components of national geographic information strategy. These are the need for coordination and the need for the development of metadata services respectively.

Given the large number of government custodians involved and the differing nature of their responsibilities with respect to the creation and maintenance of digital databases, the primary task of coordination is the development of standards and guidelines which make it possible for data held in these bases to be combined with data held in other databases in order to maximise their potential usefulness. In the process it will also be necessary to consider what incentives can be offered to custodians to share their data with new users.

A related problem is the need to take steps to provide information about what databases have been developed by public sector agencies so that potential users can find the information that they require. For this reason the development of metadata services, which provide systematic information about the spatial data holdings of different government agencies, must be recognised as the second key element of any national geographic information strategy. As Burnhill (1991, p. 16) has pointed out, 'the effective handling of metadata involves clear definitions and sensible standards and procedures'. Consequently there are many features in common between the task of coordination, which seeks to devise common standards and procedures to facilitate spatial data integration, and the standards and procedures used at a higher level to describe data holdings in metadata services.

While information has some of the features of an asset that requires effective management, it also has a number of unique qualities which need to be borne in mind in the process. These essentially reflect the issues raised in the earlier discussion of information as a resource. There are also a number of potential conflicts between the notion of information as a commodity and the notion of information as an asset which must be taken into account by governments in the development of national geographic information strategies.

GEOGRAPHIC INFORMATION AS AN INFRASTRUCTURE

The notion of geographic information as an infrastructure has much in common with that of geographic information as an asset. Nevertheless it also highlights a number of additional issues related to the management of geographic information which need to be taken into account in the development of any national geographic information strategy. The notion itself invites comparisons with other items of infrastructure such as transport networks or schools which are regarded as fundamental to the effective functioning of society. Yet the term 'infrastructure' is hard to grasp. Most dictionary definitions tend to stress the idea that infrastructure is the basic facilities that are required to meet the needs of society with respect to road and rail networks and school buildings. At the same time, however, it can also be viewed in much broader terms as referring to the human resources and the operational procedures required to run these transport networks and the teachers and teaching materials that constitute an indispensable part of the overall school system.

This broad view of the notion of infrastructure can be seen in the vision of a coordinated spatial data infrastructure for the US nation presented by the Mapping Sciences Committee of the National Academy of Sciences. This includes 'the materials, technology, and people necessary to acquire, process, store, and distribute such information to meet a wide variety of needs' (National Research Council, 1993, p. 16).

The Mapping Sciences Committee is also at pains to make clear that, while geographic information is necessary for such an infrastructure, it is not sufficient in itself: 'of equal importance are the individuals, institutions, and technological and value systems that make it a functional entity, one that serves as the basis for much of the business of the nation' (p. 17).

The Mapping Sciences Committee also points out that there are already national spatial data infrastructures in existence. What is essentially lacking is coordinated national geographic information strategies which take account of the extent to which circumstances have changed as a result of the advent of geographic information systems technology. The present national spatial data infrastructure

> is an ad hoc affair because, until very recently, no one has conceived of it or defined it as a coherent entity, and indeed it has not been very coherent or coordinated. It is not the task of the MSC . . . to create the national spatial data infrastructure. We want merely to point out its existence, identify its components and characteristics, assess the efficiency and effectiveness with which it functions to meet national needs (particularly at the federal level), and make recommendations that might make it more useful, more economical, more effective, better coordinated, and robust. (p. 17)

In these respects the notion of geographic information as an infrastructure fleshes out some of the issues that underlay the discussion of geographic information as an asset. It also reinforces the argument for the establishment of national geographic information strategies to coordinate the management of large numbers of geographic databases held by a variety of data custodians in the public and private sectors together with the need for metadata services to help potential users locate the data that they need.

The notion of infrastructure also raises questions about the relative importance that must be attached to these databases. Some of these are of greater importance than others because of the extent to which they are required for many different applications. These include the spatial referencing systems such as coordinates and administrative subdivisions which make it possible to link data from different sources, as well as the basic topographic and statistical data which provides a common reference framework for many different applications. Given the strategic importance of these core data, the arrangements that are made with respect to their supply and availability must be regarded as the third key element of any national geographic information strategy.

It must be emphasised that the creation and maintenance of core data sets as part of national geographic information strategy does not necessarily imply that they must be funded from the public purse. Usage fees are common for most other elements of infrastructure and a geographic information infrastructure is no different from these elements in this respect. It is also not an argument for governments to carry out the work directly: private or semi-private agencies can be employed to do so. However, it is an argument for regulation. One of the principal characteristics of other structural elements in the economy is that they are regulated for the common good and not left to the vagaries of the market or to exploitation by monopoly suppliers. Consequently an element of regulation and standardisation is an important component of any national geographic information strategy in practice.

CONCLUSIONS

The nature of information and geographic information has been discussed in this chapter in the context of the emerging information economy. This highlights the special significance of geographic location as a means of integrating and adding value to databases held by a large number of public and private agencies.

The nature of information and geographic information was then considered from four different conceptual standpoints: as a resource, a commodity, an asset and an infrastructure. The discussion of geographic information as a resource drew attention to some unique features which distinguish information from other resources such as land, labour and capital. These features also present problems when considering geographic information as a commodity which can be bought and sold like any other. This in turn raised questions about the role of governments with respect to geographic information and the conflicts of interest that may arise between the needs of public accountability and the desire to recover some or all of the costs of database creation and maintenance through the sale of data by government agencies. The concept of information as an asset introduced a new dimension into the discussion of the role of government with respect to geographic information. This concerns the need for governments to manage their geographic information assets in the national interest. This also introduced the concept of custodianship into the discussion. In the process it was found that two different kinds of asset management are involved: the custodianship of particular databases and the coordination and management of the national portfolio of geographic information assets. As a result, two of the three key elements of national geographic information strategy were identified: the need for coordination and the need for metadata services to enable users to find the information that they require.

The last part of the chapter dealt with the notion of geographic information as an infrastructure. Although this concept has many features in common with a notion of geographic information as an asset, it drew attention to a number of additional matters that need to be taken into account in the development of national geographic information strategies. A broad view of infrastructure was taken which highlighted the extent to which national geographic information infrastructures are not just a collection of databases but also the people, the technology and the cocktail of laws and precedents described by Rhind (1996a). It was also noted that these infrastructures already exist and that the primary task for governments is to develop coherent national geographic information strategies to build upon them in order to facilitate the use of GIS technology. In the process it was recognised that there is a need to identify the core databases that are required for large numbers of applications. Because of their overall strategic importance, these core databases constitute the final key element of any national geographic information strategy.

The main objective of this chapter was to clarify the concepts involved prior to the presentation of the findings of the four case studies. In the next four chapters the progress that has so far been made with respect to each of the three key elements of national geographic information strategies in Britain, the Netherlands, Australia and the United States will be considered in relation to the activities of the main data providers in government and the institutional context within which they operate.

ACKNOWLEDGEMENT

An earlier version of this chapter was published in a paper coauthored by the author and Robert Barr under the title 'Geographic information: a resource, a commodity, an asset or an infrastructure?' in Z. Kemp (ed) *Innovations in GIS 4*, pp. 234–48, 1997, London, Taylor & Francis.

CHAPTER THREE

Britain

The Creation of a National Digital Topographic Database

INTRODUCTION

The United Kingdom of Great Britain and Northern Ireland is a constitutional monarchy created by the union of England, Wales, Scotland and Northern Ireland. It is situated in northwest Europe and has a population of 56.7 million living on a land area of 244,000 square kilometres (see Figure 3.1). The effective union of the principality of Wales with the Kingdom of England dates back to 1301 when King Edward I was created Prince of Wales although Wales was not enfranchised until the sixteenth century. The name Great Britain was first used in 1604 after James VI of Scotland succeeded to the throne of England, but was not formally adopted until 1707 when the Parliaments of England and Scotland were combined by the Act of Union. In 1801 Ireland became part of the United Kingdom as a result of a further Act of Union, but in 1922, 26 Irish counties left the United Kingdom to form the Irish Free State, now the Republic of Ireland. From that time the territory has been referred to as the United Kingdom of Great Britain and Northern Ireland.

Given these circumstances, it is not surprising to find that the geographical coverage of government departments in the United Kingdom of Great Britain and Northern Ireland varies considerably and the main providers of geographic information are no exception to this rule. Many of these variations reflect not only regional circumstances but also differences in the legal systems of England, Wales, Scotland and Northern Ireland. For example, the main agency with responsibility for land and property in England and Wales is Her Majesty's Land Registry. Its counterpart in Scotland is Registers of Scotland. Similarly, the national mapping agency for Great Britain is Ordnance Survey. Its counterpart in Northern Ireland is the Ordnance Survey of Northern Ireland. In the case of statistical responsibilities the position is further complicated by their decentralisation to all the main government departments in England and Wales as well as their counterparts in Scotland and Northern Ireland. Furthermore, in England and Wales the registers of vital events are maintained by the Office for National Statistics which is also responsible for the decennial census of population and housing. Its counterparts in Scotland, and Northern Ireland are the General Registry Office, Scotland, and the Department of Health and Social Security in Northern Ireland. These variations are not confined to government departments. The nature and functions of local government also vary considerably between

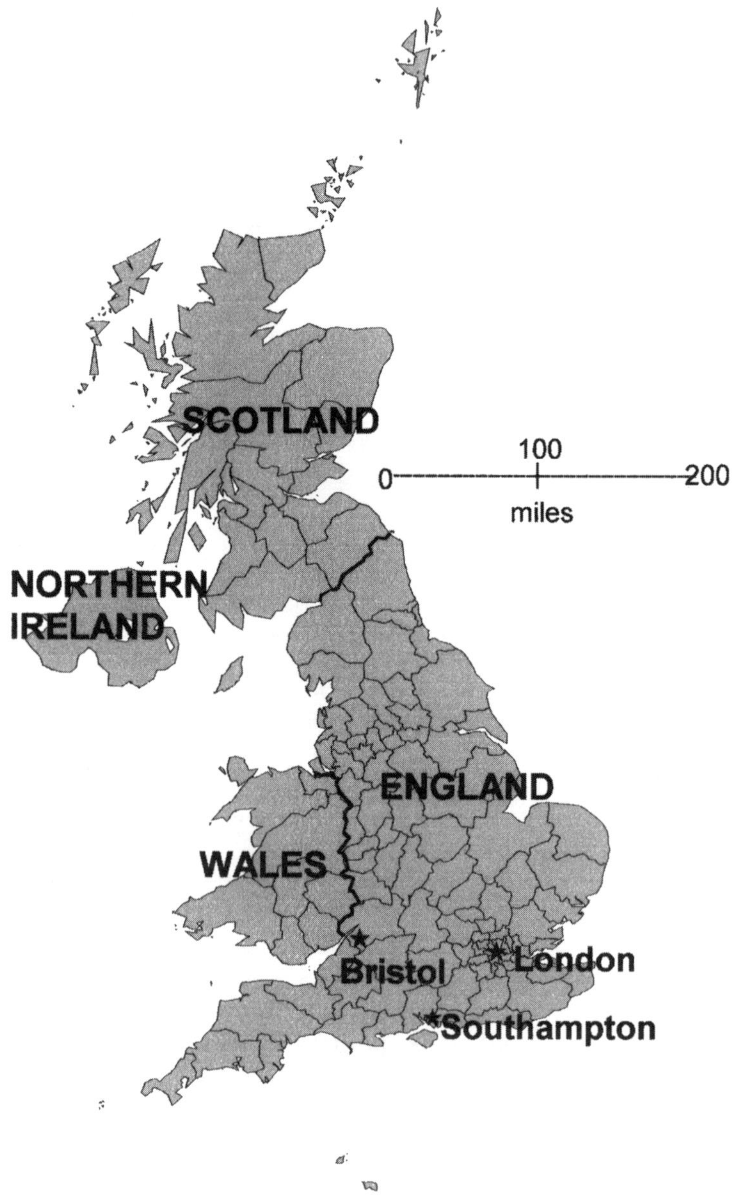

Figure 3.1 Britain.

England and Wales, Scotland and Northern Ireland. Similarly, the status of the various public utility companies varies from one part of the United Kingdom to another. For example, water provision remains in public ownership in Scotland whereas it has been privatised in England and Wales.

It is not the intention of this chapter to attempt to present a comprehensive account of the activities of all these providers of geographic information within the public sector. They will be dealt with primarily from an English perspective in the discussion that follows. Nevertheless, it is important to bear in mind not only that these variations in geographic coverage exist but also that they often embody differences in powers and responsibilities. It must also be emphasised that English practice is not necessarily the most advanced in the United Kingdom when it comes to particular geographic information activities. For example, Scotland's land registration system dates back several hundred years further than the equivalent system in England and Wales to the Register of Sasines that was set up by the Scottish Parliament in 1617.

MAIN PROVIDERS OF GEOGRAPHIC INFORMATION

Her Majesty's Land Registry

Her Majesty's Land Registry for England and Wales was set up in 1862 as a separate department responsible to the Lord Chancellor. Its main task is to provide the land title security and statutory conveyancing machinery for the transfer of property interests. It became an Executive Agency in 1990 and a Trading Fund under the Government Trading Act in 1993. It is required by statute to be self-financing and makes no call on public funds. All transactions involving property in England and Wales must be registered by the Land Registry. In 1995–6 it dealt with 4.1 million transactions, including 1.64 million new ownerships, 1.19 million new mortgages and the cancellation or variation of 1.28 million other interests in land (HM Land Registry, 1996, p. 8). In the process the number of titles to land held by the Land Registry increased from 15.5 million at the end of March 1995 to 15.99 million at the corresponding date in 1996. As a result of these activities the Land Registry generated an income of £210 million (US$315 million) in 1995–6 and produced an operating surplus of nearly £17 million (US$25 million) during the same period (HM Land Registry, 1996, p. 37).

In the last few years the Land Registry has made considerable progress in computerising these records and plans to complete the project by 1998. By the end of March 1996 over 12.8 million out of the 15.9 million titles had already been computerised (HM Land Registry, 1996, p. 12) but there are some regional variations in the extent of computerisation. For example, less than 80 percent of the titles in the East Midlands region had been computerised at this stage as against a 100 percent computerisation in most of the northern and southwestern regions.

The registers of title to land became open to public access in 1990. With over 80 percent of all titles currently held accessible on line, users are increasingly taking advantage of the direct access and telephone search services offered by the Land Registry. The direct line service permits any credit card holder with an appropriate terminal to have on-line access to the register, whereas the telephone search services enable a credit card holder to lodge applications by telephone for searches of the index map with respect to land charges.

Not all titles to land in Britain are held in the Land Registry. In the absence of compulsory registration of land titles, its records are limited mainly to land parcels where transactions have taken place since the Land Registration Act was passed in 1925.

Ordnance Survey

Ordnance Survey of Great Britain was founded in 1791. Since May 1990 it has been an Executive Agency with considerable autonomy regarding the conduct of its own affairs. 'It is responsible for the official, definitive surveying and topographic mapping of Great

Britain. It is a government department funded by Parliamentary vote. The responsible minister is the Secretary of State for the Environment' (Ordnance Survey, 1995, p. 5). To carry out these tasks Ordnance Survey employs about 2,000 staff at its headquarters in Southampton and its network of field offices throughout the country. Its expenditure in 1995/6 was £85.6 million (about US$130 million) (Ordnance Survey, 1996a, p. 34). Nearly 80 percent of this expenditure was covered by the sale of products and services.

The core activity of Ordnance Survey is to maintain the national topographic database in the form that meets current and future demands. The basic parameters of this database were established over a century ago in 1863 when it was agreed that all urban areas in Britain should be mapped at the 1:500 scale (later adjusted to 1:1,250 scale because of the costs involved), rural areas to the 1:2,500 scale and the remaining mountain and moorland areas to the 1:10,000 scale (McMaster, 1991, p. 11). Ordnance Survey completed the digitisation of all 57,799 1:1,250 map sheets, 166,877 1:2,500 map sheets and 4,040 1:10,000 mountain and moorland map sheets in May 1995. As a result it now provides an accurate up-to-date topographic database for the whole country (Rhind, 1996b, p. 20). Some examples of the different kinds of map that are available from Ordnance Survey can be found on Plates 1–4.

In addition to this database Ordnance Survey has developed a wide range of digital products derived from this and other databases. The main features of some of the most important of these are summarised in Table 3.1. From this it can be seen that they include the ADDRESS-POINT product developed in conjunction with the Royal Mail which uniquely identifies and precisely locates each of the 25 million residential, business and public postal addresses in Great Britain, the OSCAR (Ordnance Survey Centre Alignment of Roads) range of products related to street centre line information, and many others. In developing these projects Ordnance Survey works closely with both public and private sector partners.

Ordnance Survey also works closely with other public agencies in relation to the supply of geographic information (Rhind, 1995b). In a few cases these connections are specified by law. The Land Registry, for example, is obliged by statute to record its data holdings on Ordnance Survey maps. In its recent Framework Document, the Ordnance Survey also announced plans to create a National Geospatial Data Framework by linking its national topographical database to other spatial data held by government departments and agencies with a view to providing a 'one-stop shop' for customers requiring the government's spatial data (Ordnance Survey, 1995, p. 5). These plans will be considered in more detail later in this chapter in the context of the national geographic information strategies.

The Framework Document also makes clear that the government requires the Ordnance Survey to discharge its responsibilities 'while maximising the recovery of its costs in meeting customer demand' (Ordnance Survey, 1995, p. 5). The need to recover some of its costs is nothing new. Nearly 20 years ago the then secretary of state for the environment, Peter Shore, gave Ordnance Survey a 40 percent cost recovery target. However, by the early 1990s this had increased to 80 percent notwithstanding the essential contribution that its activities make to the national interest in many fields. Consequently it is not surprising to find that one objective of the framework document is 'to agree a National Interest Mapping Contract with government services and activities undertaken in the national interest' (Ordnance Survey, 1995, p. 5).

Over the last five years Ordnance Survey has changed dramatically. The extent of these changes is evident from the following statement by its director general, David Rhind (1996b, p. 20):

> Over the last few years the Ordnance Survey has reformed itself. The national address database and the road-centre-lines database did not exist in 1992, nor did the Ordnance Survey carry out consultancy or work on data products with commercial sector partners to any great extent. The Ordnance Survey is now organised into business units trading in an internal market. It is increasingly customer-driven and sees itself as much more than a traditional national mapping organisation.

Consequently, its current mission statement is 'to be customers' first choice for mapping today and tomorrow' (Ordnance Survey, 1996b).

Table 3.1 National products introduced by Ordnance Survey since 1992

Product	Source scale	Number of maps/records	When first completed	Comments
OSCAR: road-centre lines plus road names	1:1,250 to 1:10,000	0.5 million km of road	1994	Covers every driveable road in Britain. Updated every 6 months. Available at 3 levels of sophistication.
ADDRESS-POINT	N/A	25 million records stored	1995	National Address database with 0.1m coordinates on each address. Updated every 6 months.
1:50,000-scale colour raster	1:50,000	812 tiles	1994	Created in partnership with private sector.
1:10,000-scale black & white raster	1:10,000	10,556 maps	1994	Created in partnership with private sector.
ED-LINE	1:10,000	109,670 enumeration districts	1992	Boundaries of all population census areas in England and Wales. Produced in partnership with private sector.
Meridian	Various	805 tiles	1996	Designed as a topological vector database and conflated from OSCAR and other OS data inputs. A 'mid-range' product.
Land-Form PROFILE	1:10,000	10,556 maps	1996 (contours)	The National Height Model in the form of 5m and 10m contours and a Digital Terrain Model consisting of height values on a 10m grid.

Source: Rhind (1996b, p. 21)

Office for National Statistics

The United Kingdom is unusual in that it has a decentralised government statistical service which comprises the statistical divisions of all the main government departments as well as the Welsh, Scottish and Northern Ireland offices. Nevertheless, most of the main data collection and survey activities have been concentrated in two agencies, the Central Statistical Office and the Office of Population Censuses and Surveys. Since 1 April 1996 these two agencies have been merged to form a single agency called the Office for National Statistics which is accountable to the Chancellor of the Exchequer. The head of the Office for National Statistics is also the director of the Government Statistical Service and the registrar general for England and Wales. The main reason for this merger is the need to give greater coherence and compatibility in government statistics. Consequently, in addition to taking over the statistical responsibilities previously carried out by these two agencies, the new agency has been charged with establishing a new central database of key economic and social statistics drawn from the whole range of statistics produced by government to common classifications, definitions and standards (OPCS, 1995, p. 2).

The Central Statistical Office (CSO) was set up in 1941 following the concerns expressed by the prime minister, Winston Churchill, about the use of different statistical data by government departments. The main focus of its work has subsequently been the compilation of the national accounts and it also provides the coordination required for the successful operation of a decentralised statistical service. The Central Statistical Office became an Executive Agency in 1991 and it is envisaged that its successor will have an expanded role in this capacity.

Work on labour market statistics was transferred to the Central Statistical Office from the Employment Department in 1995. This includes the National On-line Manpower Information System (NOMIS) which is maintained under contract by the University of Durham. This provides subscribers with direct access to official government statistics on population, employment, unemployment, and resources down to the smallest geographical area for which they are available (Blakemore, 1991).

The Office of Population Censuses and Surveys (OPCS) was created in 1970 by the merger of the General Registry Office which had been in existence since 1837 and the Government Social Survey which came into being in 1941. Prior to the merger with the Central Statistical Office the head of OPCS was accountable directly to the secretary of state for health. The basic tasks of OPCS are to secure the registration of vital events such as births, marriages and deaths and to provide high quality demographic, social and medical information and analysis. The OPCS is also responsible for the organisation of the decennial census of population and housing in England and Wales and carries out surveys for other government departments and public bodies. In April 1994 over 1,800 staff were employed by the OPCS in 10 divisions spread over three sites in London, Titchfield and Southport (OPCS, 1994, p. 9).

Unlike the CSO, there has always been a very strong geographic dimension to the work of the OPCS. As a result this agency has worked closely with other key players in the geographic information field in the planning of its decennial censuses and the development of geographical information products from these censuses (see Coombes, 1995, for a review of current practice). The OPCS also maintains the Central Postcode Directory. This contains records of each of the 1.7 million unit postcodes in the United Kingdom together with a National Grid reference for one address within each postcode area and the OPCS codes for the administrative areas within which it lies.

THE INSTITUTIONAL CONTEXT

Disseminating Geographic Information

Traditional 'paper and ink' funding arrangements remained in place in Britain until relatively recently but data providers are now expected to operate in a more commercial way as a result of a combination of factors, including the following:

- a wider programme of reassessing the role of government, with emphasis on privatisation, deregulation and market testing;
- pressures on government to constrain spending; and
- increasing demand for data and higher quality datasets, partly as a result of improvements in computer technology, which also adds to the financial pressure on information providers (Coopers and Lybrand, 1996, p. 19).

With these considerations in mind, the government launched its Next Steps programme in 1988. The aim of this programme is 'to deliver services more effectively and efficiently, within available resources, for the benefit of taxpayers, customers and staff' (HM Treasury, 1989, para. 2.2). By the end of 1995 some 109 Executive Agencies had been created and it has been estimated that 60 percent of the Home Civil Service now works in agencies or other organisations that operate on Next Steps lines (Chancellor of the Duchy of Lancaster, 1996). These include the main government providers of geographic information such as HM Land Registry, Ordnance Survey and the Central Statistical Office. Each of these Executive Agencies is given financial, efficiency and customer service targets so that its performance over time can be closely monitored. According to Coopers and Lybrand (1996, p. 20), 'these targets have become more demanding over time, reflecting the need to ensure that limited resources are used with maximum effectiveness and efficiency'.

More specifically with respect to information, the Department of Trade and Industry launched its Tradeable Information Initiative in 1986 'to make as much government held information as possible available for the private sector to turn into electronic services' (DTI, 1990). Government departments were asked to review their information holdings to identify data that could be made available and guidelines for charging for such information were provided by the Department of Trade and Industry. These distinguish between two categories of tradeable information. In cases where an established market exists for government tradeable information which they already provide, departments were asked to charge a reasonable market price. In the case of tradeable information that has not previously been exploited, it was suggested that departments made contracts initially on the basis of charging only for costs incurred over and above those associated with handling the information for their own purposes (DTI, 1990).

In addition, HM Treasury also provides guidance on the fees and charges that must be levied by government departments. These are reflected in the guidelines for charging for statistical services and products produced by the Government Statistical Service. The most important of these are the following:

> (a) The cost of collecting, processing, analysing and presenting statistical information for official use should be met from public funds.
>
> (b) Statistical information which is published for wider use, in printed or electronic form, should normally be priced to recover the full costs of making that information available in excess of that required for government use . . .
>
> (c) The full cost is the total cost of all the resources used. For statistical services, this will commonly include product development, computing, editorial, production, marketing and distribution costs and the cost of capital . . . It will also include the costs of collection, where collection was not for official use. (GSS, 1995, p. 4)

The extent to which government information providers are in a position to recover some or all of the costs of database creation and maintenance varies considerably in practice. As Table 3.2 shows, agencies such as the Central Statistical Office component of the Office for National Statistics, which collect data primarily for government economic

Table 3.2 Government information providers' expenditure, revenue, net cash costs and cost recovery

£ million	Expenditure 1994/5	Revenue 1994/5	Net cash costs	% cost recovery
CSO*	49.5	1.9	47.6	4%
OPCS* (includes research revenue)	70.0	38.0	32.0	54%
Meteorological Office (includes research revenue)	141.1	57.0	84.1	40%
British Geological Survey (includes research revenue)	40.0	24.0	16.0	60%
Hydrographic Office	37.9	22.0	15.9	70%
Ordnance Survey	74.8	58.6	16.2	78%
Registers of Scotland	29.6	31.5	(1.8)	106%
HM Land Registry	197.4	235.6	(38.2)	119%
Companies House	33.5	36.8	(3.1)	110%

* CSO and OPCS merged on 1 April 1996.
Note: These figures are all given in UK pounds. To convert to US dollars, multiply by 1.5.
Source: Coopers and Lybrand (1996, p. 33)

policy-making, are characterised by low-cost recovery. In contrast, agencies collecting data primarily for regulatory purposes such as HM Land Registry, typically achieve something close to full-cost recovery, while agencies collecting data with a national interest component such as Ordnance Survey fall somewhere between these two extremes (Coopers and Lybrand, 1996, pp. 30–40).

As its range of digital products has increased, Ordnance Survey has negotiated Service Level Agreements with some of its main groups of customers to make its data available to them on a shared purchase basis. The first of these agreements with the local authority sector was reached in March 1993. This has resulted in a marked increase in data use. As a result, 80 percent of British local authorities were using digital data in 1995 as against 20 percent in 1993 (Rhind, 1995b, p. 2). Since that time further agreements have been reached with the public utilities and a consortium of Scottish office government departments and other agencies. Such agreements are seen as being of mutual benefit to both data supplier and data user. For Ordnance Survey they reduce the costs of administering a large number of separate agreements with individual users and they also promote a more widespread diffusion of its products than would otherwise have been the case. From the standpoint of the user there are obvious financial savings as well as less uncertainty regarding the extent of possible future commitments.

Legal Protection

Matters relating to the legal protection of geographic information in Britain are largely dealt with by the Copyright, Designs and Patents Act of 1988. This covers literary and artistic works as well as information held in computer databases. To qualify for copyright protection under the Act, a work must be 'original'. Insofar as databases are concerned it is enough to show that their compilers have expended work and effort to create them (sweat of the brow) without necessarily having to put in any creative effort. In this respect British copyright law differs from that in other European countries such as the Netherlands (see Chapter 4) where a work must bear the personal stamp of its creator to be considered as 'original' with respect to copyright (AGI, 1993, p. 9). Consequently most geographic information products in Britain are covered by the provisions of copyright law provided that they involve skill and labour and have not been copied.

The person who creates the work is regarded as the initial owner of copyright. However, employers are presumed to own the copyright in the case of works created by their employees in the course of their duties and the Crown has the copyright in works produced by civil servants.

Copyright owners are allowed a fixed period in which they may exploit their rights. For computer-generated work the duration of copyright is 70 years from their creation (Larner, 1992).

The holder of Crown copyright is the Controller of Her Majesty's Stationery Office (HMSO), and the provisions of the Copyright Act insofar as they relate to government departments are administered by the copyright unit of HMSO. This deals with requests for permission to reproduce Crown copyright material. However, the administration of Crown copyright relating to Ordnance Survey products is delegated to the director general of Ordnance Survey who is also responsible for deciding the rules and terms for reproducing this material (Ordnance Survey, 1996c). Ordnance Survey digital map data are available under a variety of copyright licensing agreements. Of particular importance in this respect is the distinction made between the conditions of digital copyright licences for agreed business use within organisations and the use of Ordnance Survey data in commercial third-party products. Given that external copyright income accounts for one-third of its turnover, Ordnance Survey fiercely defends its intellectual property rights and won a High Court case regarding copyright in 1995 (Rhind, 1995b).

In addition to copyright legislation, geographic information comes under the 1984 Data Protection Act which regulates the use of digital information relating to individuals. All users are required to register and abide by its principles to ensure that personal data are obtained and processed fairly and used only for previously specified purposes (Data Protection Registrar, 1993).

It should be noted, however, that while the Data Protection Act has been hailed by some lawyers as 'exemplary of data protection legislation of broader scope' (Perritt, 1996, p. 141), it has also been criticised on the grounds that it has not been used enough. For example, the Committee of Public Accounts (1994, para. 24) concluded:

> We do not believe that the Data Protection Principles are operating as they should to safeguard the rights of individuals, nor are they policed effectively. The Registrar monitors compliance with the Principles [of the Act] only to a very limited extent. He has no audit or inspection functions nor the power to demand information from data users. Too much depends on their willingness to cooperate.

ELEMENTS OF NATIONAL GEOGRAPHIC INFORMATION STRATEGY

Coordination

The starting point for much of the present discussion about the need for a national geographic information strategy in Britain is the publication of the Report of the Committee of Enquiry on Handling Geographic Information chaired by Lord Chorley in May 1987 (Department of Environment, 1987). This built on and consolidated the findings of two earlier committees: the Ordnance Survey Review Committee chaired by Sir David Serpell which reported in 1979 (Serpell, 1979) and the survey of remote sensing and digital mapping undertaken by the House of Lords Select Committee on Science and Technology (1983). The main body of the Chorley Report is divided into two roughly equal parts. The first of these consists of three chapters reviewing recent developments in geographic information handling technology, whereas the second deals in more detail with the specific issues involved and sets out the reasoning behind the 64 recommendations made to the secretary of state for the environment.

The first part of the report sets the scene for further discussion. It reflects the committee's enthusiasm for the new technology: 'the biggest step forward in the handling of geographic information since the invention of the map' (para. 1.7), and also its concern that information technology in itself must be regarded as 'a necessary, though not sufficient condition for the take up of geographic information systems to increase rapidly' (para. 1.22). To facilitate the rapid take up of GIS, the committee argues that it will be necessary to overcome a number of important barriers to effective utilisation. Of particular importance in this respect are the need for greater user awareness and the availability of data in digital form suitable for use in particular applications.

Given this context, the committee's most important recommendations are those relating to digital topographic mapping, the linking and availability of data and the role of government. With respect to digital topographic mapping, its recommendations for an accelerated programme of digital data conversion using a simplified specification for conversion purposes are now largely a matter of historic interest, given the completion of the national digital topographic database by Ordnance Survey in 1995.

In contrast, the committee's recommendations on data availability and data linkage are still of considerable relevance. They stress the importance of maximising the use of geographical data held by government and other public sector agencies. It is vital for this purpose therefore that the agencies involved should take the necessary steps to make their data available to users, preferably in an unaggregated form, except where this is prevented by the question of confidentiality. The committee also points out that the main benefits of introducing GIS depend to a large extent on linking data sets together, and that it will be necessary to develop standard forms of geographic description to facilitate linkage based either on the coordinates of the National Grid or, in the case of socio-economic data, on postcode areas used by the Royal Mail.

Underlying the whole report is the argument that government has a central role to play in the future development of the field and that it must give a clear lead in this respect. With this in mind the committee considered a number of options before finally recommending the establishment of an independent Centre for Geographic Information with strong links to government:

1. to provide a focus and forum for common interest groups, or clubs;
2. to carry out and provide support for promotion of the use of geographic information technology . . .

3. to oversee progress and to submit proposals for developing national policy in the following areas:
   - the availability of government spatial data, the operation of data registers and arrangements for archiving of permanent data;
   - the development of locational referencing, standard spatial units for holding and releasing data, the operation of the post code system and the development of data exchange standards (cartographic and non cartographic);
   - the assessment of education and training needs and provision of opportunities to meet them; and
   - the identification of R&D needs and priorities, including advice to government on bids for R&D funds (para. 10.22).

Subsequent developments with respect to the development of strategies for promoting data availability and facilitating data linkage will be covered later in this section. As far as the proposals for a national Centre for Geographic Information are concerned, the government's response was generally negative. It argued that it was better to encourage existing organisations to expand their range of activities rather than set up a new organisation (see Masser, 1988, and Rhind and Mounsey, 1989, for a full discussion of these issues).

Nevertheless, a new organisation, the Association for Geographic Information (AGI), was set up in January 1989 with help from the government. According to its current director, the basic mission of the AGI is to spread the benefits of geographic information and to help all users and vendors of GIS (Leslie, 1994). The work of the AGI is concentrated around three main activities: informing, influencing and acting. Informing activities include its publications, seminars and conference. Influencing activities involve liaison with government agencies and other organisations to encourage the greater use of geographic information, while acting activities include various special projects initiated by the AGI itself. The most important of these from the standpoint of the development of a national geographic information strategy are the work that has been undertaken on the development of standards both in the United Kingdom and in Europe, and the round-table discussions initiated by the AGI with bodies like Ordnance Survey on data charging policy and the Office of Population Censuses and Surveys on problems regarding the availability of 1991 census data and planning for the 2001 census. In a recent series of meetings with key government organisations, the round-table format has also been used to explore some of the key issues that need to be resolved with respect to the further development of a national geographic information infrastructure such as data availability, marketing, legal protection and standards (AGI, 1995).

The AGI is a cross between a learned society and a trade organisation which tries to represent the interests of both its individual and its corporate members who include the main suppliers of geographic information and its users (Leslie, 1994). The AGI has over 700 members from central and local government, the private sector, the utilities and academia. The subscriptions of these members provide the core funding for its activities. They are governed by a council elected by the members and serviced by a small secretariat of two full-time staff and a part-time director, based in offices at the Royal Institution of Chartered Surveyors in Westminster. The AGI also represents Britain's interests on the European Umbrella Organisation for Geographic Information (EUROGI).

Another outcome of the Chorley Report is the official working group that was set up to follow up those of its recommendations that required cooperation among government departments (Oliver, 1996a, p. 26). This group was re-launched in 1993 as the Interdepartmental Group on Geographic Information (IGGI). This regularly brings together representatives from over 30 government agencies covering a great diversity of applications (see Table 3.3).

Table 3.3 Membership of IGGI on 1 January 1996

ADAS
Ministry of Agriculture, Fisheries and Food
British Library
CCTA (Government Centre for Information Systems)
Central Statistical Office
Department for Education and Employment
English Nature
Department of the Environment
English Heritage
ETSU (Energy Technology Support Unit)
Forestry Commission
General Register Office for Scotland
Department of Health
Home Office
Hydrographic Office (Ministry of Defence)
HM Land Registry
Lord Chancellor's Department
Meteorological Office (Ministry of Defence)
Military Survey (Ministry of Defence)
Department of National Heritage
National Rivers Authority
Office of Fair Trading
Office of Population Censuses and Surveys
Ordnance Survey of Northern Ireland
Ordnance Survey
Scottish Office
Department of Social Security
Department of Trade and Industry
Department of Transport
HM Treasury
Valuation Office Agency
Welsh Office

IGGI seeks to:

- provide a forum for government departments to consider and develop a common view on issues affecting or affected by geographic information;
- facilitate the effective use, both within and outside central government, of geographic information held by government departments;
- consider barriers to realising the fullest potential of government held geographic information and take practical steps to overcome them.

The chairman and secretary of the IGGI are based in the Planning and Land Use Statistics division of the Department of the Environment. This division also services the Information Development Liaison Group (IDLG) which is the main central/local government statistics forum.

The IGGI has been very active in the debate regarding the development of a national geographic information strategy. It also sees itself as having a key role to play in the debate between the government and the wider geographic information community. With this in mind, it has worked closely with the AGI in the planning and organisation of the government round-table meetings (Oliver, 1996b).

A good example of the spirit of collaboration within IGGI is the project that it has commissioned on the UK standard geographic base (Massingham, 1996). This aims to provide a framework for standard area names and codes and a standard set of core spatial references. It is anticipated that the core spatial references and referencing system developed as a result of this project will be implemented as an official standard, taking into account the work of the relevant national and international bodies.

Core Data

It will be clear from the previous discussion that many of the elements of a national core database for geographic information are already in place in Britain. These include the national topographic database maintained by Ordnance Survey as well as the land and property information held by the Land Registry and the socio-economic data held by the Office for National Statistics. Consequently the main issues are generally perceived in Britain in terms of the linking of these data sets rather than their creation, as is the case in many other countries. Of special interest in this connection are two complementary initiatives that are taking place at the present time. The first of these involves a pilot project to evaluate the feasibility of setting up a National Land Information Service in England and Wales to link data holdings from several sources. This project also provides a useful opportunity to test the recently agreed BS7666 *Land and Property Gazetteer* standards in operation. The second project outlines a vision of a National Geospatial Data Framework which goes considerably further than the data holdings linked in the previous pilot project.

National Land Information Service (NLIS)

Many groups have argued for some time that special priority must be given to the establishment of a National Land Information Service (see, for example, Dale, 1994). As a result, a joint working group of the Department of the Environment, Ordnance Survey, the Land Registry, the Valuation Office and the Local Government Management Board was set up in 1994 to undertake a live pilot project to explore the feasibility of establishing a national land information system. The project itself covered two postal districts in Bristol containing about 30,000 properties located in over 700 streets. The basic task for those involved was to link together data held by Ordnance Survey, the Land Registry, the Valuation Office and the City of Bristol itself, using the unique property reference numbers (UPRN) contained in the *Land and Property Gazetteer* developed by the Local Government Management Board with the encouragement of the AGI, and embodied in BS7666 which were published in three parts during 1993 and 1994. These define the industry standards for indexing street information and indexing land and property data, and the address structure required by the other two indexes respectively (Cushnie, 1994).

The system developed for this purpose operates on the Land Registry's main frame but provides on-line access from a single terminal to databases held by Ordnance Survey in Southampton, the Valuation Office in Worthing, the Land Registry in Plymouth and Bristol City Council in Bristol. During the trial phase of this project a high matching rate of over 95 percent was achieved for residential properties which indicates the potential for even higher matching as data structures are further refined (Smith, 1996).

The findings of the NLIS pilot project demonstrate the potential advantages to be derived for users, especially legal users requiring information about property ownership and details regarding valuation and planning constraints relating to particular properties, from linking key data sets. In the light of this experience the NLIS steering group has commissioned a live conveyancing project which is being undertaken in conjunction with the further development of a *Land and Property Gazetteer* for Bristol, and a feasibility study of the business case for NLIS services. The terms of reference for this feasibility study are as follows:

- establish the potential demand for NLIS services;
- identify ownership and funding for NLIS, including its operational management;
- identify likely charging frameworks for customers and suppliers;
- establish priorities and timescales for implementation.

(Smith and Goodwin, 1996, p. 2.16.1)

National Geospatial Data Framework (NGDF)

The 1995 Ordnance Survey Framework Document included among its objectives plans to create a National Geospatial Data Framework by linking its national topographic database to other spatial data such as those held by other government departments. A subsequent report from the Ordnance Survey gives some indication of the form that this National Geospatial Data Framework might take (Nansen, Smith and Davey, 1996). The authors argue that the American concept of a national spatial data infrastructure is not applicable to Britain because of the relative autonomy of British government departments, the continuous pressures on them to reduce costs and/or raise revenue and the lack of freedom of information legislation. On the other hand, there has been formal encouragement from British ministers for government departments to work together and the secretary of state for the environment has asked Ordnance Survey to be proactive in promoting greater cooperation. With this in mind, the authors set out a vision of the National Geospatial Data Framework as a virtual database which is not the property of any one organisation but 'the totality of many individual datasets collected and held separately by many different organisations' (p. 4).

It is argued that such a loose structure can only be made to work if certain standards are accepted and adhered to by all participants. Consequently a data set will qualify for inclusion in the National Geospatial Data Framework only if it meets the following conditions:

- the data set has been created and maintained to certain defined common standards.
- the characteristics of the data set are defined in a common way and made generally available through a standard and easily accessed channel.
- the data set is accessible and defined in publicised terms.
- it is possible to link this and other data sets together through a set of standard tools.

(p. 4)

Given the initiatives already under way in Britain, it is felt that there is no need for a 'big bang' approach with its consequent requirement for large-scale funding to achieve such a vision.

> It is the Ordnance Survey view that the national geospatial data framework will be created gradually by the process of data providers entering into partnerships – joining the club – when the time is appropriate for them to do so. This will be facilitated by the creation of the overriding data architecture described above and a technical delivery architecture to integrate data sets and to deliver the applications. (p. 7)

What is needed, therefore, is to bring together all these initiatives into a combined programme of research and development in order to refine the concept of a National Geospatial Data Framework. Following discussions between Ordnance Survey and a wide range of interested parties, the following management structure was set up at the end of 1996 to implement the NGDF:

- an NGDF Board, led by Ordnance Survey to decide policy and commit resources.
- an NGDF Advisory Council, led by AGI, as a strong channel for inputting the views of the geospatial community.
- an NGDF Task Force, working to the Board. (Davey and Murray, 1996, p. 6)

Metadata

The Spatial Information Enquiry Service (SINES) is a simple metadata service which has been administered by Ordnance Survey since 1994. It is a database which contains details of almost 600 spatially referenced data sets held by over 40 government departments and related bodies. These details include the following:

- the title of the data set.
- the purpose for which the data were collected.
- the method and source of data collection.
- the time period covered by the data set.
- the geographic area covered.
- the data items.
- spatial references used.
- base map usage.
- system/software used to store the data.
- data availability.
- contact point for further information. (Garnsworthy and Hadley, 1994)

SINES can be accessed by telephone, fax, email or directly through the World Wide Web. To find out what data sources exist on a particular topic or for a particular area, users can search the database using various key words. In its current form SINES provides a useful overview of what spatial information is held by government departments and also gives contact points for each data set so that potential users can obtain further information regarding their availability if they need to do so.

EVALUATION

The findings of the above analysis highlight the diversity of interests involved in geographic information in Britain within the public sector. In addition to these interests there is an even greater diversity of private sector interests ranging from hardware and software vendors to information providers and from public policy pressure groups to the academic research community. This diversity is reflected in the view that the body set up to represent these interests, the Association for Geographic Information, has to be regarded as a cross between a learned society and a trade organisation.

The analysis also draws attention to the fact that in most cases there is no single agency responsible for various types of geographic information in the United Kingdom as a whole because of the special circumstances governing the United Kingdom. Consequently it is necessary, for example, to consult two different agencies (Ordnance Survey and the Ordnance Survey of Northern Ireland) when topographic data for the whole country is required and three different agencies (ONS, the General Registry Office, Scotland, and the Department of Health and Social Security in Northern Ireland) to bring together census of population data for the United Kingdom. Given these difficulties and the relative size of the constituent elements, it is also not surprising to find that much of the debate has been dominated by English interests and that the differences between these and the other members of the United Kingdom are not always fully appreciated by outsiders.

These variations present a number of problems with respect to access to data held by government departments which are exacerbated by the different practices between departments with respect to the dissemination of geographic information. As a result, users seeking the same information from the equivalent agencies in the constituent members of the United Kingdom may get different answers from them when they request information. The findings of the AGI/government round table highlight the extent of 'these perceived inconsistencies in policies relating to the availability of and access to geographic information' (Oliver, 1996b, p. 1.3.2). They also draw attention to the lack of awareness in many government agencies of the value of geographic information and the unwillingness of many departments to release such information due to lack of resources to tackle issues such as product definition, pricing, copyright, liability, privacy and support for the development of marketing services (AGI, 1995, p. 1).

Access to information held by government departments is not only a problem for the private sector. A recent study of geographic information system use in government departments and non-departmental public bodies (Hookham, 1995, p. 20) found that 'the availability of digital information, at an affordable price, was the biggest single barrier to public sector use of GIS'.

Coopers and Lybrand (1996, p. 25) have also noted the tensions that exist in government policy between the pressures on data providers, who are urged to take a more commercial view, and the restrictions placed on their participation in the development of value-added products and services. At the same time they point out: 'There may also be an increasing conflict in some areas between the financial objectives set for government information providers and the pressure for greater openness in the provision of information.' This has prompted observations such as the following from some outside commentators: 'Pricing policy in the UK is arbitrary. It is unclear whether the government wants to recover costs, make a profit or stimulate the private sector' (Policy Studies Institute, 1995, p. 66).

A good example of these tensions and conflicts is the government's response to the European Union Directive on the freedom of access to environmental information (CEC, 1990) which requires public agencies in member states to make environmental data available to the public on demand at a reasonable cost. However, the term 'reasonable cost' has still to be defined, although the UK regulations implementing the Directive state that this 'may include provision for the imposition of a charge on any person in respect of the costs reasonably attributable to the supply of information to that person' (Department of the Environment, 1992).

Given these conflicts and tensions, it can be argued that there is a need to regulate the main providers of geographic information on the grounds that they have some of the characteristics of natural monopolies. However, it can also be argued that these providers are regulated in some way at the present time through ministers who are accountable to

Parliament. As a result of their analysis of the costs and benefits involved in regulating geographic information provision in the United Kingdom, Coopers and Lybrand (1996, p. 8) conclude that the case for price control regulation is relatively weak for the following reasons:

- the small size of the information providers in relation to, say, the utilities;
- the difficulties in defining suitable output metrics;
- the absence of strong monopoly power in some markets; and
- the sophistication of the key users of these products and services.

Copyright and data protection issues also need much greater attention than has hitherto been the case, given the nature of the changes that are currently taking place with effect to both the *de jure* and *de facto* positions. The findings of the AGI/government round table draw attention to the need for clarification on these matters with respect to the provisions of current legislation (AGI, 1995, p. 2). However, at the same time it is also necessary to take account of potential impacts of developments at the European level which may alter this *de jure* position. For example, the new EU directive on legal protection of databases (CEC, 1996) has important consequences for copyright provision and may bring Britain more into line with other European countries in this respect. It is also necessary, however, to take account of *de facto* developments with respect to geographic database creation as a result of the introduction of new imaging technologies which provide alternatives to conventional topographic maps for certain applications.

It can be seen from the analysis that the Chorley Report has had a considerable impact on the development of national geographic information strategies in Britain. It has played a major role in raising awareness of the issues involved and has also stimulated the establishment of the AGI as a forum for the national geographic information community and the creation of the predecessor to IGGI to coordinate geographic information activities within government departments. The recommendations of the Chorley Report also prompted the acceleration of the Ordnance Survey national topographic database digitisation programme which was completed in 1995.

Despite this, in some key respects of national geographic information strategy there has been less progress. Notwithstanding its achievements in informing, influencing and acting, the AGI has yet to take a leading position with respect to the formulation and coordination of national geographic information strategy. Similarly, IGGI relies heavily on voluntary collaboration between its members. Nevertheless it has built up a spirit of collaboration within government departments in a relatively short time. Despite these achievements, however, there is no British equivalent with the same political support and the same powers of the US Federal Geographic Data Committee which has played such an important role in the development and implementation of its National Spatial Data Infrastructure (see Chapter 6).

In the absence of such a body, Ordnance Survey has played a key role in the development of national geographic information strategies. It should also be noted that the director general of Ordnance Survey is the official adviser to the UK government on all survey, mapping and GIS-related matters. For this reason its recent shift in emphasis from the establishment of a national topographic database to a vision of a National Geospatial Data Framework which links data sets held by other agencies must be regarded as an important move in the development of a national geographic information strategy.

The merger of the Central Statistical Office and the Office of Population Censuses and Surveys to form the new Office for National Statistics is likely to have important consequences for national information policies as a whole in future years. Of particular importance in this respect is likely to be the establishment of a new central database of key economic and social statistics to give greater confidence and compatibility in government statistics.

Nevertheless, the findings of the analysis show that there is a great deal of interest in the development of national geographic information strategies in Britain among the key players and that Britain is in some respects ahead of many other countries now that its core digital national topographic database is in place. Consequently, the key players have been able to focus their attention on measures such as the National Geospatial Data Framework to link other data sets to this core database rather than the creation of the core database itself. The results of the National Land Information Service pilot project are also encouraging. This shows that the technical problems involved in linking data in different centres to create a virtual database can be overcome. In both cases, however, the next stage of these projects is likely to be crucial from the standpoint of the further development of national geographic information strategies. The vision of the National Geospatial Data Framework has still to be fleshed out into concrete proposals. Similarly the business case for a National Land Information Service has still to be developed and accepted by government.

CHAPTER FOUR

The Netherlands

The Emergence of a National Geographic Information Strategy

INTRODUCTION

The Netherlands is a constitutional monarchy located on the northwest coast of Europe (see Figure 4.1). As its name suggests, the country is flat and low lying with more than half of its land area below sea level. It is also one of the most densely populated countries in the world with 15.25 million people living on only 40,800 square kilometres of land. The history of the modern Netherlands began in 1579 when the Seven United Provinces which lie to the north of the River Maas declared their independence from the Spanish Netherlands. In 1815 two further provinces, North Brabant and Limburg, became part of the Netherlands and a further province, Flevoland, has been created since the Second World War as a result of the reclamation of a large part of the former Zuiderzee.

From the point of view of geographic information supply, most of the main tasks in the public sector are centralised in the Netherlands. Land titles registration and large-scale cadastral mapping is the responsibility of the Cadastre, small-scale mapping is in the hands of the Topografische Dienst and the collection of statistics is the task of Statistics Netherlands. However, responsibility for real estate valuation is delegated to the 650 municipalities, as is the task of maintaining the municipal population registers. The activities of these key providers will be described in the next section of this chapter.

MAIN PROVIDERS OF GEOGRAPHIC INFORMATION

The Cadastre

Since 1832 the Cadastre has maintained a public register of land titles for the Netherlands as a whole. Its current records contain information about 7 million land parcels owned by 3.5 million owners. Each year it records around 350,000 notarial deeds, 340,000 new mortgages, 260,000 terminations of mortgages and 110,000 subdivisions of parcels. Information about these parcels is maintained on 30,000 cadastral maps which are updated continuously (Besemer, 1994, p. 7).

A complex organisation has evolved to carry out these tasks employing 2,200 people with an annual turnover of over DFL 450 million (US$300 million) in 1995 (Kadaster,

Figure 4.1 The Netherlands.

1996, p. 38). About 200 of these staff are based at its headquarters in Apeldoorn. The remainder are distributed among 15 regional offices in the 12 provinces of the Netherlands and its IT department.

Details regarding land titles are contained in the Automated Cadastral Register (AKR) and information about the land parcels is held in a Land Survey and Cartographic Information System (LKI). The Register is now fully automated and about three-quarters of the LKI is currently available in digital form.

Unlike many other cadastral organisations in Europe, the Dutch Cadastre combines both the registration and land survey activities in a single organisation. It formed part of the Ministry of Finance until 1974 when it was transferred to the Ministry of Housing, Spatial Planning and the Environment (VROM). In 1994 it became an independent administrative body (Zelfstandige Bestuurs Orgaan). Under the terms of its special foundation law, the secretary of state for the Ministry of Housing, Spatial Planning and the Environment must approve its long-term policy plans and the fees that it charges, and he also has the right to be kept informed. In every other respect the Cadastre operates as an independent organisation except that it is required by law to recover its operational costs and to return any profits that it makes to its customers by cutting its charges. Because of efficiency gains and cuts in staff since 1994, as well as a fast-moving property market, the Cadastre has been able to cut its fees by an average of 45 percent (van der Molen, 1996, p. 35).

The current philosophy of the Cadastre can be summarised as follows:

> We may run the Dutch Cadastre as a commercial organisation but we are aware that it is not commercial in the normal sense. We exert a certain monopoly insofar as the legal part is concerned. Providing legal security in the land market is not and may not be, a commercial activity, it is a public matter. However, our mission is to exert such a monopoly in a customer driven way. We perform professional market research and make decisions on better services, new products, better delivery times. (van der Molen, 1996, p. 35)

For some years the Cadastre has marketed a number of products derived from its databases as well as participating in the production of a large-scale base map for the Netherlands together with the utility companies and the municipalities (see the section on core data). These include a variety of postcode and grid square data products. In April 1996 the Cadastre also set up a separate company to develop a new postal address coordinate product for all 7 million postal addresses in the Netherlands.

The Topografische Dienst

Although modern topographic mapping began in the Netherlands during the first decades of the nineteenth century, the Topografische Dienst in its present form dates from only 1932. It is part of the Ministry of Defence but its employees are all civilians. Originally located in Delft, it moved to Emmen in the northeast Netherlands in 1984.

The primary duty of the Topografische Dienst is to maintain the national topographic base at scales of 1:10,000 and above in the interests of national security. By comparison with the Cadastre it is small in size with less than 160 employees and a budget of about DFL 20 million (about US$12.5 million) in 1994 (Opie, 1995, p. xvii). Because of its position within the Ministry of Defence, the Topografische Dienst has limited flexibility of operation, and is also being increasingly required to cover part of its operational costs through sales of analogue and digital products, and specialised services. For the 1996 financial year it has been given a cost recovery target which is nearly 25 percent of its overall budget of DFL 21.8 million (US$14 million) and it is envisaged that this will rise to nearly 50 percent by 2000.

A series of 1:25,000 scale maps has been available for the Netherlands since 1865 and the Topografische Dienst began the production of a 1:10,000 series in 1952. It currently produces a wide range of hard copy cartographic products ranging from the 1:10,000 to the 1:500,000 scale including reproductions of historical maps as well as current series. These are mainly sold by the Topografische Dienst itself through the postal system although they are also available from a limited number of booksellers.

Over the last 10 years the Topografische Dienst has become increasingly involved in the production of digital data products at various scales and in various formats. Its digital vector products include the TOP10vector (1:10,000), TOP50vector (1:50,000) and TOP250vector (1:250,000) as well as 1:10,000 scale street-centre line data. It also sells, through a separate company, 1:25,000 and 1:250,000 raster products derived from its analogue maps. Of particular importance from the standpoint of national geographic information strategy in the Netherlands is the work that is currently under way to create a 1:10,000 scale core database for the Netherlands. This will be described in some detail later in this chapter.

Statistics Netherlands (Centraal Bureau voor de Statistiek)

The collection of statistics in the Netherlands is centralised at the national level under Statistics Netherlands which was set up by royal decree in 1899. Statistics Netherlands is part of the Ministry of Economic Affairs from which it receives most of its funding but remains essentially independent in its statistical activities.

Statistics Netherlands was substantially restructured in 1993 into a flatter organisation consisting of a number of divisions which have been given a great deal of autonomy (Statistics Netherlands, 1995). In the new structure there are four divisions devoted to the collection and integration of statistical data relating respectively to agriculture, industry and environment; trade, transport and services; socio-cultural statistics; and socio-economic statistics. There are also divisions devoted to data collection, office services and research and development as well as a new statistical division which has been given the task of integrating the information collected by other divisions and the presentation and marketing of this information. This division also has a general responsibility for regional statistics and recently set up its own GIS unit to expand its data integration and presentation capabilities.

An important theme underlying this restructuring is the need to encourage more widespread use of the wide range of data collected by Statistics Netherlands and to exploit the potential of modern information technology. This vision is encapsulated in the subtitle of its corporate strategy document which can be translated as 'from a factory of figures to a node on the electronic highway'. This highlights an important change in emphasis from data collection to data integration as well as stressing the importance of linking Statistics Netherlands data to data generated by other suppliers.

Within this general context Statistics Netherlands (SN) is developing a number of products using GIS technology. A good example of these projects is the land use cover statistics produced by the agriculture division. A small group was set up in 1989 to produce a comprehensive inventory of land use cover statistics using data derived from the Topografische Dienst 1:10,000 scale hard copy maps and aerial photography. The initial inventory was completed in 1994 and the group is currently updating its database using 1993 data. The work of the group has featured in SN publications (see, for example, Visser and Lengkeek, 1995) and has also attracted interest from the Ministry of Agriculture, Nature and Fisheries (LNV), the National Institute for Public Health and the Environment (RIVM) and the Ministry of Housing, Spatial Planning and the Environment (VROM) as well as private sector agencies. It has also proved a useful test bed for evaluating the comprehensiveness of the 1:10,000 scale digital base that is being developed by the Topografische Dienst (Lengkeek, 1996).

Other Public Data Suppliers

In addition to these products two administrative registers that are maintained in the Netherlands are of considerable importance. The population registers, which are maintained by the municipalities (Gemeentelijke Basis Administratie (GBA)), contain details about every individual in the country. The quality and accuracy of these registers is such that they have replaced the decennial census since 1971 (see Redfern, 1987, pp. 187–97 for a detailed discussion of the background to this decision). Aggregate and disaggregate anonymised information from those registers is made available at the national level by SN. Since 1995 the personal identifiers used in these registers have been linked to the land titles records held by the Cadastre.

The business registers maintained by the Chambers of Commerce perform a similar function for private firms in the Netherlands. In addition to the products derived from this source by SN, the Dutch Association of Chambers of Commerce has set up its own organisation to market information obtained from them.

INSTITUTIONAL CONTEXT

Disseminating Geographic Information

There is no specific policy for disseminating geographic information collected by the public sector in the Netherlands. The Ministry of Internal Affairs (BIZA) has overall responsibility for coordinating information policy but has few powers in real terms to persuade other ministries to adopt particular standards. As a result, policy is decided at the ministry level or even at the section level within a ministry rather than at the national level (Policy Studies Institute, 1995, p. 42).

Despite this, there are growing pressures on public agencies to recover some or all of the costs of data collection and distribution. This is particularly evident in the geographic information field because of its value to end users and also because of the availability of much of this data in digital form. Generally the preferred mode of operation throughout government agencies in the Netherlands is to exploit their information assets at arm's length rather than by dealing with third parties directly (Policy Studies Institute, 1995, p. 42).

This is reflected to some extent in the dissemination activities of all three key players. The Cadastre is required by law to recover its costs of operation but at the same time it is forbidden to make a profit from its core administrative activities. Consequently any surplus that it makes from these activities must be returned to its customers by way of reduced charges. Under the terms of its Foundation Act the Cadastre is allowed to carry out additional activities provided that they are not subsidised by its core business and are approved by its supervisory board and the secretary of state. In April 1996 it set up a separate company to develop a new postal address coordinate data product for the whole of the Netherlands.

The Topografische Dienst receives core funding from the Ministry of Defence. Nevertheless it is being increasingly required to recover part of its operating costs through the sale of its analogue and digital products and specialised services. Its target for 1996–7 is nearly a quarter of its entire budget and it is anticipated that this proportion will increase to nearly half by 2000. The Topografische Dienst sells digital data to both public and private sector agencies at various rates according to whether the data are to be used for internal purposes or for the development of value-added products for resale. In each case it agrees a contract with the users which sets out the uses of the data that are permitted. The Topografische Dienst handles the majority of sales itself. Only the one-off 1:25,000 and 1:250,000 raster versions of its analogue maps are marketed through a commercial organisation.

Statistics Netherlands is also under pressure to increase the use of its data by more effective marketing. Generally, however, its prices cover only the costs of distribution rather than the full costs of data collection and distribution. Like many other government agencies in the Netherlands it also makes a distinction between commercial and other users, charging the former four times the end-user cost plus royalties (Policy Studies Institute, 1995, p. 43).

As a result of its restructuring in 1993 a new integration and presentation division has been set up specifically to increase the use of SN data. It has been given a target of increasing usage by 10 percent a year up to 2000. It is increasing the number of its joint ventures with commercial value-added resellers and also making some SN data available through the Internet. In addition, it has recently set up a small GIS unit to assist in the integration of GIS data with data from other suppliers and the presentation of this data to users.

Legal Protection

A recent study of copyright on geographic information in the Netherlands for the National Council of Geographic Information (van Eechoud, 1995) discusses the problems that may arise in connection with the copyrighting of geographic information products under Dutch law. The basic arguments underlying the findings of the study can be summarised as follows:

- The Dutch Copyright Act of 1912 (Auteurswet) grants exclusive rights of reproduction and publication to the author of a work of literature, art or science.
- The interpretation of a work under this law is broad. It includes geographic information products so long as these have been laid down in a form susceptible to the senses.
- However, copyright law only exists in original works. Case law requires these works to have a character of their own and/or bear the personal stamp of their creator.
- The amount of time and effort that goes into the creation of a work (sweat of the brow) is not considered relevant for copyright under Dutch law.
- Facts are not protected in their own right because they are not original. However, a collection of facts such as a geographic information product can enjoy the protection of copyright provided that it expresses the personal vision of the author with respect to their selection, classification and retrieval.
- Many geographic information products are likely to experience difficulties meeting this criterion. For example, problems will occur in cases where standard classifications are used which are not original to the work. Furthermore, because of the use of modern information technology, the design of the output from such a work lies to a large extent in the hands of the user rather than the creator.
- Further problems may arise in determining copyright where geographic information products involve the integration of data collected by several agencies and/or the creation of the database is undertaken as a collaborative venture.

Article 10 of the Dutch constitution protects the privacy of citizens in general terms and the Personal Records Act specifies the contents of and access to personal records required for the protection of privacy. The Act states:

> a personal record kept by a governmental body may only contain such information as is relevant and necessary to serve the purpose for which that record is kept. Supplying personal records to third parties is allowed only when this supply is a direct consequence of the purpose for which the record is kept, when there are legal grounds for the supply, or when the persons registered have given permission. (van der Molen, 1994, p. 5)

However, there are a number of important exceptions to these principles with respect to the accessibility of geographic information held by public bodies. For example, the 1992

Cadastre Act (Kadasterwet) requires that the information contained in its real estate registers must be made accessible to the public.

ELEMENTS OF NATIONAL GEOGRAPHIC INFORMATION STRATEGY

Coordination

The Dutch Council for Real Estate Information (Ravi) was set up in 1984. Its primary role was to advise the minister for housing, spatial planning and the environment on matters relating to the operations of the Cadastre. In 1990 the Council was asked to review its position in the light of the increasing computerisation of real estate and geographic information services and the growing autonomy of the Cadastre. With these considerations in mind, the Council carried out an extensive survey of geographic information usage in the Netherlands and produced a master plan setting out its future role as a National Council for Geographic Information (Ravi, 1992a, b and c).

The main findings of the Ravi survey are summarised in Figure 4.2. This shows the key roles played by the Cadastre, the Topografische Dienst and Statistics Netherlands in the provision of geographic information to both public and private sector users. At the same time the analysis highlights the extent to which the various directorates of the Ministry of Housing, Spatial Planning and the Environment, together with the two levels of local government, occupy central positions as users of information from a wide variety of agencies. In contrast, the utility companies and the water boards, together with the notaries, are large users of geographic information but these tend to be drawn from a relatively limited number of sources. The diagram also demonstrates the degree to which individual households in the Netherlands are both receivers and creators of geographic information.

The Ravi master plan sets out a vision of a national geographic information strategy for the Netherlands: 'In short, the proper development of a National Information Infrastructure requires a well thought out policy, an adequate administrative organisation, and the intensive coordination of all the involved parties' (Ravi, 1992a, p. 7, author's translation). Two different types of coordination task are defined in the master plan. The first of these is termed 'field coordination'. This is required to produce a better balance between users of information on the one side and suppliers of information on the other. The findings of the survey showed that, in the past, there had been too much emphasis on the supply side and demand-side factors had not received enough attention. This resulted in an overemphasis on detailed operational information at the expense of the needs of policy-making. As a result, the links between these two components have not been adequately developed. This supply-side emphasis also meant that not enough attention has been given to future demands for geographic information, a factor of considerable importance given the growing use of information technology in planning and decision-making and the increasing availability of geographic information systems.

The second type of coordination task is termed 'core data coordination'. The findings of the Ravi survey show the intensity and the complexity of real estate and geographic information flows within the Netherlands among a large number of different agencies. This information can be seen in overall terms as a very large communications system. However, the findings of the survey also indicate that the effective operation of the system as a whole has been hindered by a number of important bottlenecks. There were many examples of failures of communication between the parties involved which resulted in

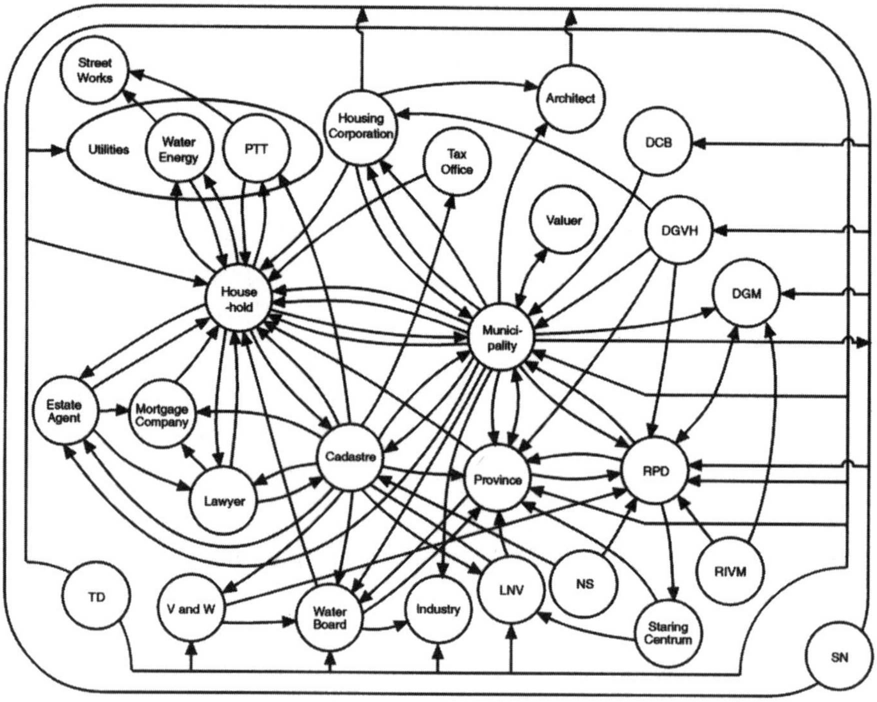

Key

DCB	Buildings	Directorates of the Ministry of Housing, Spatial Planning and the Environment (VROM)
DGVH	Housing	
DGM	Environment	
RPD	Spatial Planning	
LNV	Ministry of Agriculture, Nature and Fisheries	
NS	State railways	
RIVM	National Institute for Public Health and the Environment	
V and W	Ministry of Transport, Public Works and Water Management	
TD	Topografische Dienst	
SN	Statistics Netherlands	

Figure 4.2 Information flows between providers and suppliers in the Netherlands. *Source*: Ravi, 1992a.

the unnecessary duplication of effort on data collection and management. Many respondents also expressed their concern about the split between the large-scale information provided by the Cadastre and the small-scale topographic information provided by the Topografische Dienst. Consequently, it was argued that there is a need for the development of a digital map base somewhere between these positions at around the 1:10,000 scale. This would be particularly useful for policy-related applications which utilise GIS. As a result, it was proposed that a digital version of the 1:10,000 scale map base should be produced by the Topografische Dienst as a matter of national urgency. This was seen in terms of a standardised product which could be made available for the whole of the Netherlands within three years at low cost (see below).

With these considerations in mind the Ravi was restructured as a national consultative body for geographic information on 3 December 1993. At the time of restructuring it was made clear that there was no room for a top-down organisational structure as there was already a great deal of intragovernment cooperation in the field of geographic information. 'Because of the changing function of the advisory body from advice to consultation, and because of the general discussion in parliament concerning the role and place of formal advisory bodies, it became necessary to set up a consultative body for geographic information' (Ravi, 1995a, p. 1). The board of the new Ravi includes most of the main data providers and users in the public sector. These include the Cadastre, the Topografische Dienst and Statistics Netherlands, together with representatives from various groups within the Ministry of Housing, Spatial Planning and the Environment (VROM), the Survey Department of the Ministry of Transport, Public Works and Water Management (V and W), the National Institute for Public Health and the Environment (RIVM) and the Department of Land Development and Soil Mapping (Staring Centrum), as well as representatives from the Association of Provincial Agencies (IPO), the consultative group of the public utilities companies, the Royal Association of Civil Law Notaries, and the Association of Waterboards. The Association of Dutch Municipalities (VNG) also supports the Ravi by contributing to the final costs of the projects but is not a full member of the board.

In summary, then:

> The new Ravi comprises all the public services and local authorities with an important role in the provision of real estate and geographic information. These organisations aim to improve the [national] spatial information infrastructure by means of cooperation and agreement, thereby forming a uniquely authoritative organisation which will, in years to come, develop its position as [the] leading neutral consultative body. (Ravi, 1995c, p. 16)

To carry out these tasks, core funding of DFL 500,000 a year (about US$350,000) is provided by the Ministry of Housing, Spatial Planning and the Environment but the other members together contribute about three times this amount to support the Council's work. This covers the costs of a small secretariat based in Amersfoort with four full-time staff. Its initial core funding extended for three years and the Ministry agreed in 1996 to continue its core funding for the Ravi's activities until 2001.

The secretary of state for housing, spatial planning and the environment is still politically responsible for coordinating the Ravi's activities as well as keeping Parliament informed about them. The relationship between the secretary of state and the Ravi is governed by a covenant which provides for regular discussions between the Ravi and the minister and the reporting of its work.

Prior to 1993 there were representatives from the private sector on the board of the Ravi. In the process of restructuring it was decided to restrict membership to public sector organisations to avoid potential conflicts of interest. Despite this, the Ravi is very much aware of the need to consult widely across the whole of the Dutch geographic information community. Consequently it set up a business platform in May 1994 to facilitate consultations with the business world and to inform the Dutch geographic information industry about its products. At the present time some 20 businesses are members of the business platform. These include most of the main hardware and software vendors in the Netherlands. Recently the Ravi has also set up an academic advisory board consisting of leading Dutch geographic information scientists to further assist it in its activities.

Despite its small size, the restructured Ravi has carried out a large number of projects during its brief life span in conjunction with both public sector agencies and private companies. It has played an important part in the establishment of the 1:10,000 scale core database (see below). It has also carried out useful work on standardisation and copyright and sees itself as the node in the European geographic information infrastructure for the Netherlands. In this respect it has played an important role in the discussions regarding European geographic information policy and matters relating to standardisation and copyright within Europe.

The Council's views on the further development of a National Geographic Information Infrastructure are set out in a paper published in November 1995 to promote discussion between politicians, government agencies, the business world, and scientific and professional groups (Ravi, 1995c). It defines the National Geographic Information Infrastructure as 'a collection of policy, data sets, standards, technology (hardware, software and electronic communications) and knowledge providing a user with the geographic information needed to carry out a task' (p. 6). The discussion paper makes a distinction between core data and thematic data. It is argued that the existence of certain core data sets is vital to the functioning of society because of their widespread application. These form the basic cornerstones of a national geographic information infrastructure and their supply must be guaranteed in the national interest. For this reason the primary task is to improve the cohesion between these data sets to maximise their general usefulness. Examples of these core data sets include the cadastral register itself, the large-scale base map of the Netherlands and the 1:10,000 digital database that is being developed by the Topografische Dienst. With respect to the thematic data that are required for particular applications in fields such as spatial planning, transport, the environment or water management, it is argued that questions of harmonisation must be tackled in a rather different way by bringing the users themselves together to determine how the data should be managed. A key instrument for facilitating the accessibility of existing data is the National Geographic Information Clearing House project (see below).

The impact of recent initiatives designed to develop the information market in the private sector is clearly apparent in the discussion paper. For example, 'properly coordinated, existing geographic information in the Netherlands can create new economic opportunities. By utilising geographic information, existing tasks, services and products can be improved and new information services developed' (p. 4).

Core Data

Two core digital topographic databases are under development in the Netherlands at the present time: the large-scale base map of the Netherlands (GBKN) and the 1:10,000 core database. The main features of these are described below. In addition, a number of other initiatives are under way to facilitate the integration of existing databases. The most important of these was the decision in 1995 to use the municipal population register personal data identifiers in the land title records held by the Cadastre (Ravi, 1995b).

The Large-scale Base Map of the Netherlands Project (GBKN)

The GBKN (Grootschalige Basiskaart Nederland) project began in 1975 with an agreement between the Cadastre and some of the main users of large-scale map data in the Netherlands (the utility companies and the municipalities) to develop a large-scale map series mainly at the 1:1,000 scale. However, progress during the following years was variable due to the high costs of the project and the unwillingness of the Cadastre to

Table 4.1 Progress of the GBKN project at 1 January 1996

Province	Area (ha)	% Completion 1.1.1996	
		(ha)	(%)
Groningen	240,043	26,791	11.2
Friesland	371,400	302,550	81.5
Drenthe	268,351	162,489	60.6
Overijssel	357,820	183,000	51.1
Flevoland	153,855	153,855	100.0
Gelderland	514,749	257,500	50.0
Utrecht	143,971	124,416	86.4
North-Holland	290,323	113,881	39.2
South-Holland	309,630	248,916	80.4
Zeeland	178,738	140,000	78.3
North-Brabant	513,209	513,209	100.0
Limburg	222,002	129,182	58.2
TOTAL	3,564,091	2,355,789	66.1

Source: (Mom, 1996, p. 15)

shoulder most of the costs of a project that gave it only limited benefits. To deal with these problems a new agreement was signed on 11 November 1992 between the Cadastre, Dutch Telecom (PTT Telecom BV), the Association of Energy Suppliers (Vereniging Energie-Ned), the Association for Water Suppliers (Vereniging van Exploitation van Waterleidingbedrijven in Nederland) and the Association of Dutch Municipalities (Vereniging Nederlandse Gemeenten) (GBKN, 1992). Under the terms of this agreement, it was agreed that:

- The participating parties would complete a large-scale base map for the whole Netherlands within 10 years at an estimated cost of DFL 350 million (US$240 million).
- Twenty percent of the costs would be met by the Cadastre, 60 percent by the three utility organisations and the remaining 20 percent by the municipalities.
- The implementation of the project was placed in the hands of a number of provincial public private partnerships (PPPs) consisting of the Cadastre, Dutch Telecom and, wherever possible, the regional utility companies and the municipalities.
- The project as a whole was to be directed by a steering committee made up of representatives of each of the signatories to the agreement, and its overall administration was placed in the hands of a small secretariat based in Amersfoort.

Since the agreement was signed, considerable progress has been made in the conversion of data to digital format. By 1 January 1996 it was estimated that large-scale digital data were available for 66.1 percent of the entire country (Mom, 1996, p. 15). Given the rate of current developments the whole project may be completed by the end of 1997.

The current position with respect to data conversion is summarised in Table 4.1. From this it can be seen that there are considerable differences in the levels of completion between provinces. For example, whereas the province of North Brabant is fully digitised, only 11.2 percent of the province of Groningen falls into this category.

It should be noted, however, that what is emerging is not one single large-scale database for the Netherlands but a number of separate databases prepared by the provincial

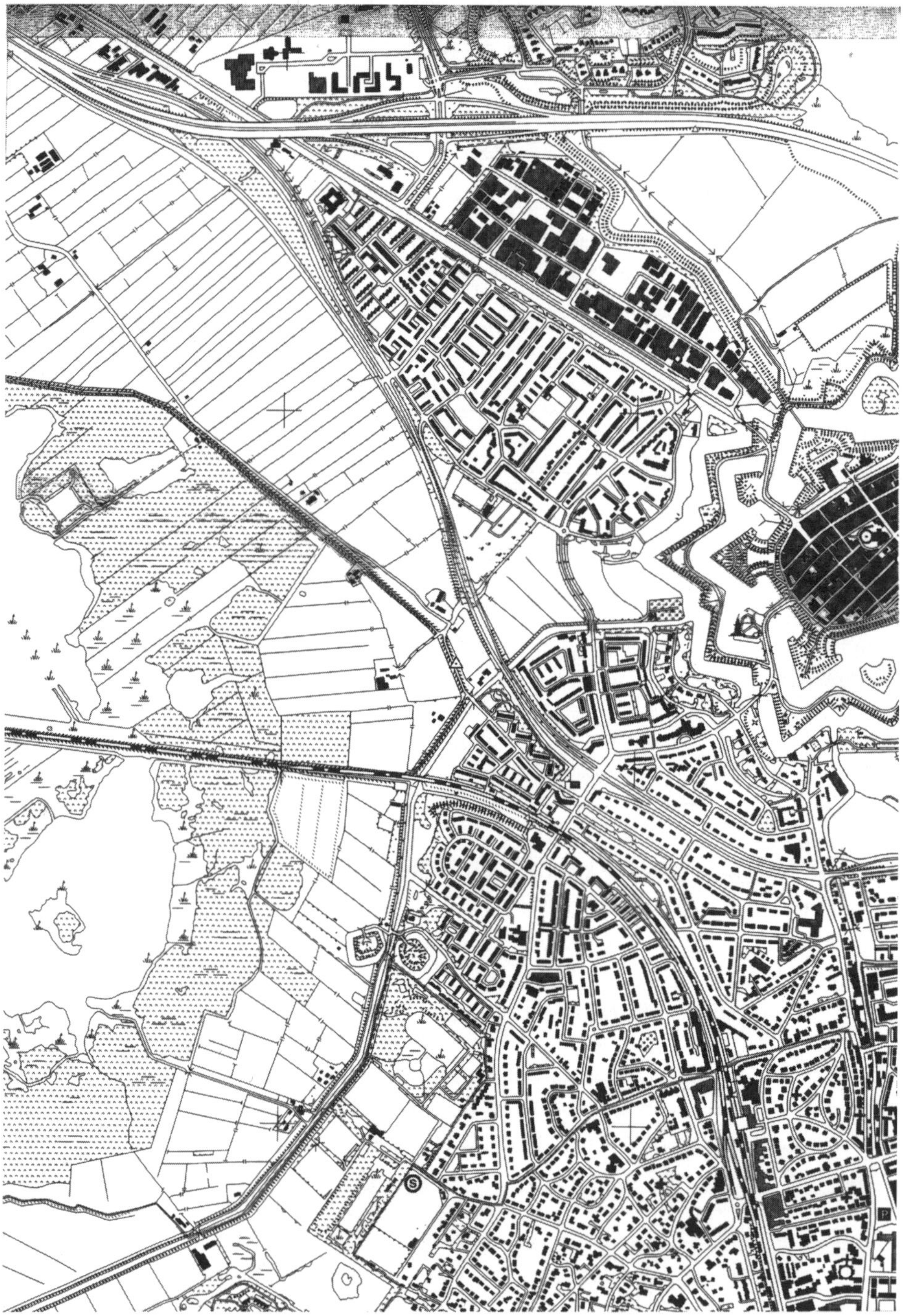

Figure 4.3 Example of Top10 vector database output for Naarden near Amsterdam. Copyright © 1996 Topografische Dienst, Emmen.

public private partnerships to slightly varying specifications within an overall set of guidelines. The position is further complicated by the insistence of many of the large municipalities to carry out the registration themselves outside these partnerships. It has already been accepted that 47 of these municipalities satisfy the necessary requirements for this purpose and the position of a further 17 municipalities has yet to be resolved (Mom, 1996, p. 17). Consequently there could be over 75 separate subsets of the GBKN when the project is completed.

Generally, this does not present problems for the main partners in the project given that their management needs are largely limited to specific localities, but it raises considerable problems for other potential users such as the police who wish to use the GBKN for their emergency planning operations. Of particular importance in this respect is the need for greater coordination and clearer guidance with respect to the ownership of this data and the terms on which they can be made available to other users.

The 1:10,000 core database

One of the main recommendations of the Ravi master plan was the need to develop a 1:10,000 digital core topographic database within three years at low cost (Ravi, 1992b). As a result, a feasibility study was carried out by a working party consisting of representatives from most of the main potential users as well as the Topografische Dienst itself. The findings of this study (Ravi, 1994) can be summarised as follows:

- It indicated the extent of the demand that existed for such data among big public sector users such as the Ministry of Housing, Spatial Planning and the Environment (VROM), the Ministry of Agriculture, Nature and Fisheries (LNV), the Ministry of Transport, Public Works and Water Management (V and W), Statistics Netherlands (SN) and the Association of Water Boards.
- It also showed that these users were willing to contribute towards the costs of the maintenance of the digital database at the 1:10,000 scale. The Ministry of Defence accepted the initial costs (at least DFL 50 million or US$35 million) and 50 percent of the maintenance. It was estimated that the costs for the maintenance would be around DFL 8.5 million (US$5.5 million) a year.
- It demonstrated that the project could be completed by the end of 1997.
- It also estimated that the collective use of the database would result in savings of around DFL 40 million for the big users in the public sector.

In the light of these findings the Topografische Dienst was ordered by the Ministry of Defence to undertake the project. The project is being undertaken in two stages (van Asperen, 1996). The first of these involves the conversion of the existing analogue 1:10,000 maps to digital maps and at the same time the implementation of a full revision of these data. At the beginning of 1996, 60 percent of the Netherlands was covered by the digital TOP10vector structure and it is envisaged that all 675 sheets will be available in digital format by the end of 1997. As a temporary solution all maps have been available in a digital form without revision since the end of 1995. Figure 4.3 contains an extract from the TOP10vector database for part of the city of Naarden near Amsterdam. The second stage is the introduction of a new four-year updating cycle for the whole of the Netherlands. As noted above, the product is being marketed by the Topografische Dienst to both public and private sector users.

Metadata

The National Clearing House for Geographic Information Project began at the Ravi in March 1995. It builds upon the experience of a number of metadata initiatives by various organisations in the Netherlands. The main objective of the Clearing House project is to develop a uniform entry point to geographic information in the Netherlands. A national metadata system of this kind is seen as fulfilling the following functions:

- better and faster access to policy information for public and private sectors.
- elimination of double work.
- forming a buying in and marketing channel for geographic data sets as half finished products or information services.
- quality improvement as a result of reporting of changes through a broad users' field.

(Ravi, 1995c, pp. 11–12)

It is intended that the National Clearing House for Geographic Information will offer something more than a basic 'yellow pages' type of catalogue of data sources. This implies developing browsing and viewing capabilities in the system and ultimately giving access to the data sets themselves.

With these considerations in mind, a survey of potential users is being carried out by the Ravi to ascertain the nature of the demand for these services. Proposals have been submitted to the Ministry of Economic Affairs for funding under the special National Action Programme which is part of the Dutch government's response to the Action Plan approved by the European Union to implement the proposals contained in the Bangemann Report to the European Commission on Europe and the Global Information Society (Tommel, 1995).

EVALUATION

The findings of the analysis indicate some of the steps that are being undertaken in the Netherlands towards the creation of a national geographic information strategy. Given the small size of the country and its high population density, it is not surprising that the management of geographic information has always been given a high priority in the Netherlands. Its significance was dramatically highlighted in January and February 1995 by the urgent need to take emergency evacuation measures on a massive scale in the face of the flood threat posed to the province of Gelderland by the exceptionally high water levels in the Maas and Rhine rivers. The position facing the emergency planners in the province on Monday 30 January 1995 has been described in the following terms:

> The first question on Monday morning 30 January was short and clear: tell us what follows from the flood threat. Then tell us what would happen if the flood occurs. Consider the impact on the various groups involved such as citizens, hazardous enterprises, agricultural units for livestock and other enterprises. Present all this in the form of a picture of the threatened areas within the province of Gelderland. Time available to find answers to these questions: less than two hours. (Akkers *et al.*, 1995, p. 9, author's translation)

As a result of development of an integrated GIS facility by the provincial government, the first results of the analysis were placed on the table less than two hours after the request for information and made a critical input into the discussions of the emergency planners responsible for managing the evacuation of large numbers of people and their goods and livestock to areas outside the threatened regions.

Colour plate section

Different map representations centred on the same area in Britain, showing different levels of generalisation in Ordnance Survey maps at different scales. Crown copyright reserved.

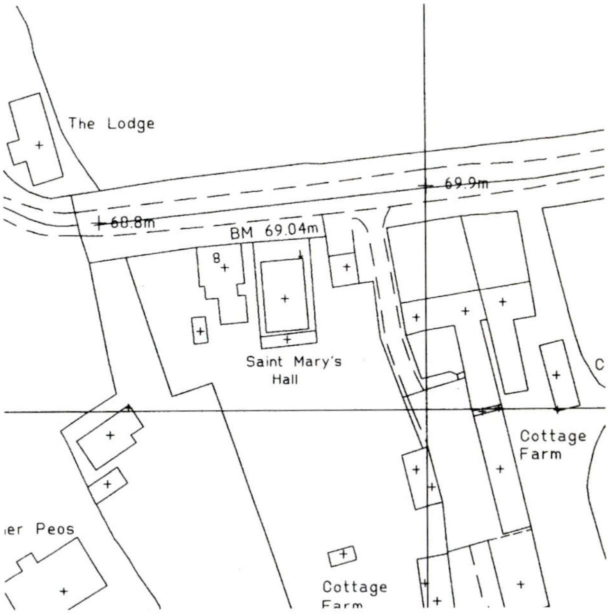

Plate 1a Landline™ data plotted at 1:1250 scale, showing house names, property seeds, pavements and road-centre lines.

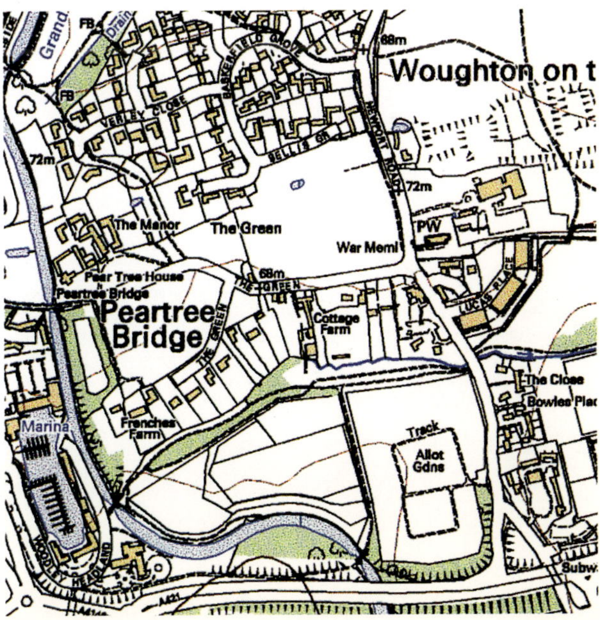

Plate 1b Landplan™ map plotted on customer request from generalised Landline data at 1:10,000 scale.

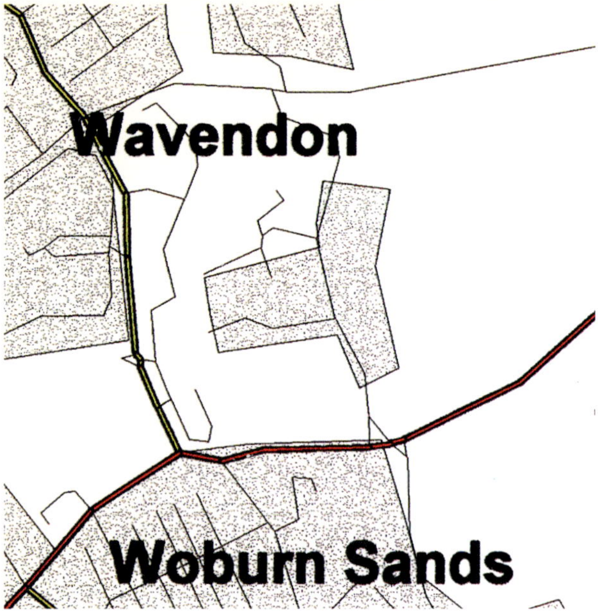

Plate 2a Meridian™ data assembled from different sources and plotted at 1:25,000 scale.

Plate 2b Pathfinder® mapping produced to a traditional specification and plotted at 1:25,000 scale.

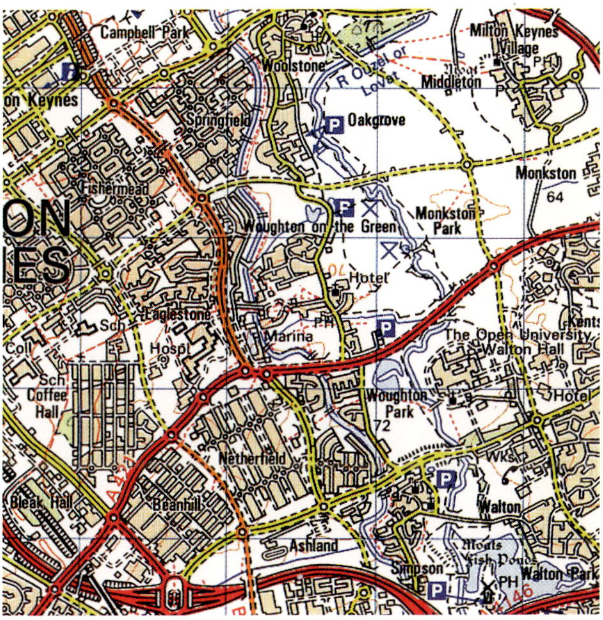

Plate 3 Landranger® data plotted and published at 1:50,000 scale.

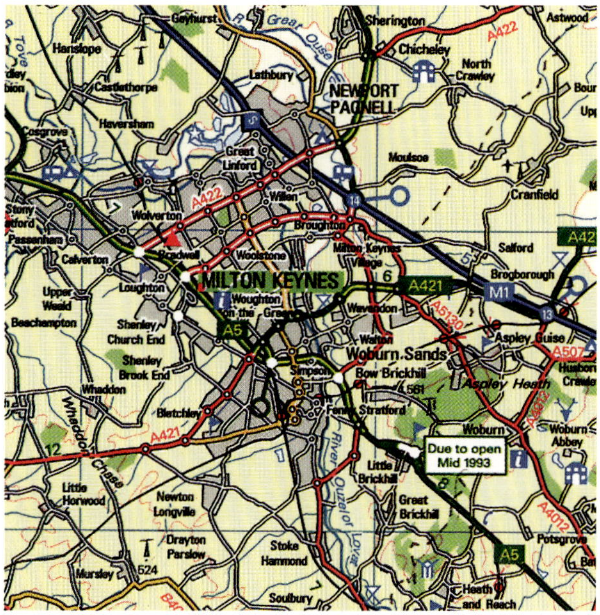

Plate 4 Travelmaster® data plotted and published at 1:250,000 scale.

Acknowledgement

All images Crown copyright reserved. Reproduced with permission of Ordnance Survey and GeoInformation International from *Framework for the World*, edited by David Rhind and published by GII in July 1997.

The findings of the above analysis also highlight the changes that are currently taking place within the public sector as a whole in the Netherlands and their implications for the development of national geographic information strategies. The consequences of these changes are particularly apparent in the restructuring of the Cadastre as an independent self-financing organisation and the reorganisation of Statistics Netherlands. Of particular importance in the case of the latter and the Topografische Dienst are the growing pressures towards the recovery of at least part of the costs of data collection and dissemination.

The extent to which these developments will restrict access to government information is not yet clear but they obviously present problems for users with limited resources. However, the rising costs of geographic information are only one of a number of factors governing the access and availability of much government-held geographic information. Other factors affecting access include the reluctance of departments to release data on the grounds that they have no mandate to do so, fears about the consequences for personal privacy, and the lack of resources for repackaging the data to make them accessible to users.

These problems are not restricted to the dissemination of geographic information to the private sector. There are also obstacles to the exchange of information between government departments in the public sector. One of the most important of these is the growing requirement for government agencies to purchase data created by other government agencies in order to carry out their public administrative duties.

There is also a need for further clarification of the legal position regarding the copyright of databases in the light of the findings of the study carried out by van Eechoud (1995, p. v). This concludes:

- In its current form copyright is not a suitable instrument for giving geographical information legal protection. Trying to extend the Copyright Act is not recommended.
- Further research into alternative regulation (particularly the EU's draft Directive on the legal protection of databases) will show the extent to which alternative protection provides a solution.

Despite these difficulties, considerable progress has been made in implementing the main recommendations contained in the 1992 Ravi master plan and the Ravi itself has completely restructured its activities to take on the role of a National Council for Geographic Information with a key part to play in the development and implementation of national geographic information strategy. In the process it has managed to maintain its position as a coordinating body for public sector agencies while broadening its operations to include private sector and research interests through its new business platform and academic advisory board.

Nevertheless, it must be borne in mind that the Ravi is a very small organisation. It should also be noted that it has no formal powers to compel public sector agencies to participate in its operations. Consequently it is very much dependent on the enthusiasm and willingness of its members to support its efforts. However, given the extent of intragovernment cooperation in the Netherlands over the last few years, this essentially pragmatic approach has paid rich dividends.

The future of the Ravi was reviewed at the end of 1995 by the secretary of state for housing, spatial planning and the environment. As a result of this review, the Ministry has agreed to continue its core funding for the Ravi's activities until 2001. Nevertheless a number of changes in its future activities seem likely. These include a greater emphasis on national geographic information policy issues such as commercialisation and legal protection as against technical issues. It is also likely that the private sector will play a greater role in its activities in the future. With this in mind, the members of Ravi's business platform prepared their own policy statement for publication in mid-1996.

Although a great deal of progress has been made over the last few years, many of the core data elements of a national geographic information strategy have yet to come into being. As a result of the split of large-scale and small-scale mapping responsibilities between the Cadastre and the Topografische Dienst, two different core data sets are in the process of creation: the large-scale base map of the Netherlands and the 1:10,000 scale topographic database. The need to harmonise these two data sets to create a single digital national topographic database must therefore be regarded as a matter of some urgency.

CHAPTER FIVE

Australia

The Commonwealth Dimension of State Geographic Information Strategies

INTRODUCTION

Australia consists of six states (New South Wales, Queensland, South Australia, Tasmania, Victoria and Western Australia) and two territories (Australian Capital Territory and Northern Territory), which together form the Commonwealth of Australia (see Figure 5.1). It is the largest island in the world with an area of over 7.5 million square kilometres which makes it a little smaller than the United States excluding Alaska and rather smaller than Europe as a whole. The history of modern Australia dates back just over 200 years to 1786 when the British government decided to settle the country. The first settlers arrived in Botany Bay in January 1788 and founded the colony of New South Wales. During the first half of the nineteenth century new colonies were also founded in Western Australia and South Australia, and Queensland and Victoria were created as a result of the subdivision of New South Wales in the 1850s. These colonies together with Tasmania were largely autonomous until the Commonwealth of Australia came into being in 1901.

Under the Australian administrative system, land matters are largely dealt with at the state and territory level while the collection of statistics is a Commonwealth responsibility. However, both are highly centralised at their respective levels. The first section of this chapter describes the land administration activities carried out at the state and territory level and then discusses the activities of the Australian Surveying and Land Information Group and the Australian Bureau of Statistics. The organisations that have emerged to coordinate these key players at the Commonwealth level will be dealt with later in the section on national geographical information strategies.

MAIN PROVIDERS OF GEOGRAPHIC INFORMATION

State Land Administration

Land administration in Australia is a state or territory responsibility. Consequently there are eight separate systems of land titles registration in Australia and eight separate surveying and mapping systems even though there are many similarities between them. Many of the principles and concepts underlying Australian cadastral systems are based

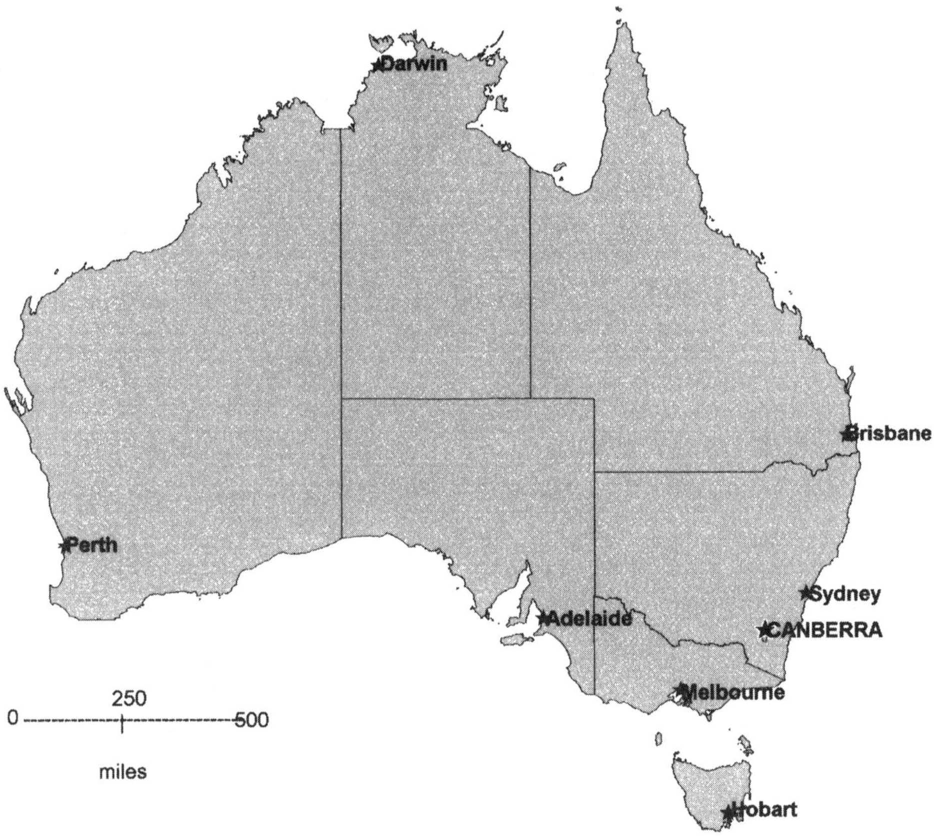

Figure 5.1 Australia.

on English common law (Williamson and Enemark, 1996). In the process the system of land transfer that evolved under English common law has been adapted to the needs of a vast newly developing country.

In the first instance it was assumed that all land was the property of the Crown. The primary task of the surveyor general in each of the states and territories was therefore to deal with the alienation of Crown land and to maintain the county- and parish-level maps. Once land was alienated, the administration of titles was typically the responsibility of the registrar general or the director of the land titles office in each state or territory. Consequently a dual system grew up in Australia which distinguished between Crown lands and alienated or private lands.

A distinctive feature of Australian land administration is the conveyancing system introduced by Robert Richard Torrens in the mid-1850s in South Australia which is used in all the states and territories. Under this system the state government issues a certificate of title for each parcel of land and guarantees its validity. The original certificate is typically held by the registrar general in the state land titles office and the owner has a duplicate copy. All subsequent dealings relating to the land parcel are recorded in a standard form on this certificate. If the parcel is subdivided into several lots, a new certificate is issued for each lot.

In this way the Torrens system replaced the complex land search and conveyancing procedures used in many other countries by a single act. However, it should be noted that

Table 5.1 Key features of the states' and territories' digital cadastral databases

	Area (thousands sq. km)	1993 Population (million)	Number of parcels (million)	Completion date
New South Wales	801.6	6.02	3.15	1993
Queensland	1727.2	3.55	1.2	1993
South Australia	984.0	1.47	0.8	1986
Tasmania	67.8	0.47	0.3	1998*
Victoria	227.6	4.47	2.3	1994
Western Australia	2525.5	1.69	0.86	1990
Australian Capital Territory	2.4	0.30	0.1	1989
Northern Territory	1346.2	0.17	0.05	1989

* Estimated.
Source: Wan and Williamson (1995a, p. 50)

it applied only to new sales and it was not until 1945 that the compulsory registration of titles was introduced in South Australia (Department of Lands, 1986). Even now several other states retain vestiges of the pre-Torrens system (Williamson and Enemark, 1996).

An important consequence of the land administration systems that have developed in Australia is the high degree of centralisation that exists with respect to the handling of cadastral records at the state and territory level. Given the parallel development of surveying and land titles administration it is not surprising, therefore, to find that a great deal of effort has been devoted to the computerisation of these activities and that most states have already established digital cadastral databases. Table 5.1 summarises some of the key features of these efforts. From this it can be seen that there are considerable variations in both land area and population between states. Whereas the largest state, Western Australia, covers an area that is more than 10 times that of the United Kingdom and over a quarter of the whole of the United States, the smallest territory, the Australian Capital Territory (ACT) which surrounds Canberra, covers an area that is less than one-thousandth of that of Western Australia. Similarly, the most heavily populated state, New South Wales, has 30 times the population of the Northern Territory.

These variations are reflected in the number of land parcels involved in each case. New South Wales has over 3 million land parcels whereas there are only 50,000 in the Northern Territory. Table 5.1 also shows the completion date of the digital cadastral database (DCDB) in each state and territory. From this it can be seen that all the mainland states and territories have completed their digital cadastral databases. Tasmania plans to finish this stage in 1998. In most cases the task has been carried out since 1990 but South Australia has already more than 10 years' experience in maintaining such a database.

As might be expected, the scales used in the process vary considerably between urban, rural and outback areas. New South Wales, for example, maps urban areas at scales from 1:500 to 1:4,000 and rural areas at scales between 1:25,000 and 1:100,000. The Australian Capital Territory, on the other hand, creates its DCDB by entering subdivision calculations with a realistic accuracy of ±0.03 metre for each parcel corner (Wan and Williamson, 1995a, p. 42).

Given these circumstances, it is not surprising to find that there is a long tradition of coordination of geographic information activities in all the states and territories. To illustrate their operation in practice the strategies developed by South Australia and Victoria are described below. These represent contrasting positions with respect to geographic information coordination.

South Australia

South Australia lies on the south central coast of Australia. Its modern settlement history dates from 1836 when it became an autonomous colony. Its land area is nearly 1 million square kilometres but two-thirds of its vast interior has insufficient rainfall to support any significant human or animal populations. As a result, most of its 1.47 million inhabitants live close to the coast, two-thirds of these in its capital, Adelaide.

The coordination of land information in South Australia goes back as far as 1974 when the state government commissioned a land information study to consider the related data needs of the then departments of lands, valuation, engineering and water supply, highways and education (Barnes, 1995, p. 99). The current body with responsibility for coordinating land-related information activities in South Australia is the South Australia Land Information Council. This council is appointed by the State Cabinet and operates primarily at the executive level. It consists of representatives of the main government departments involved in land information at the state level. Another committee, the Land Information Systems Committee (LISCOM), functions at the operational level in the state. Prior to 1992 its membership was restricted to government agencies but since that time its membership has been enlarged to include representatives from the three local universities, local government and the Australasian Urban and Regional Information Systems Association (AURISA). LISCOM is serviced by the Department of Environment and Natural Resources.

Figure 5.2 shows that the South Australian land information system is built around four functional nodes:

- Legal/fiscal: the Land Ownership and Tenure System (LOTS) integrates the data needed for the collection of government revenue with information based on land ownership. It also provides an enquiry network which gives on-line access to Commonwealth, state and local government agencies as well as the private sector and the general public.
- Geographic: this includes the digital cadastral database (DCDB), the cadastral survey data and topographic data as well as a subdivision proposal layer which is used by the utility companies for forward planning purposes.
- Environmental: this node contains a number of loosely coupled natural resources databases, the most important of which is the environmental database.
- Socio-economic: this node has still to be fully implemented (see Barnes, 1995, p. 102) but it is envisaged that it will hold a wide range of statistical data from local, state and federal government sources.

Victoria

The state of Victoria is second to New South Wales in terms of population and production even though it is the smallest in size of the five mainland Australian states with a land area of only 227,000 square kilometres (Table 5.1). It became a self-governing state in 1855 after separating from New South Wales in 1851. Nearly three-quarters of its 4.5 million inhabitants live in its capital, Melbourne.

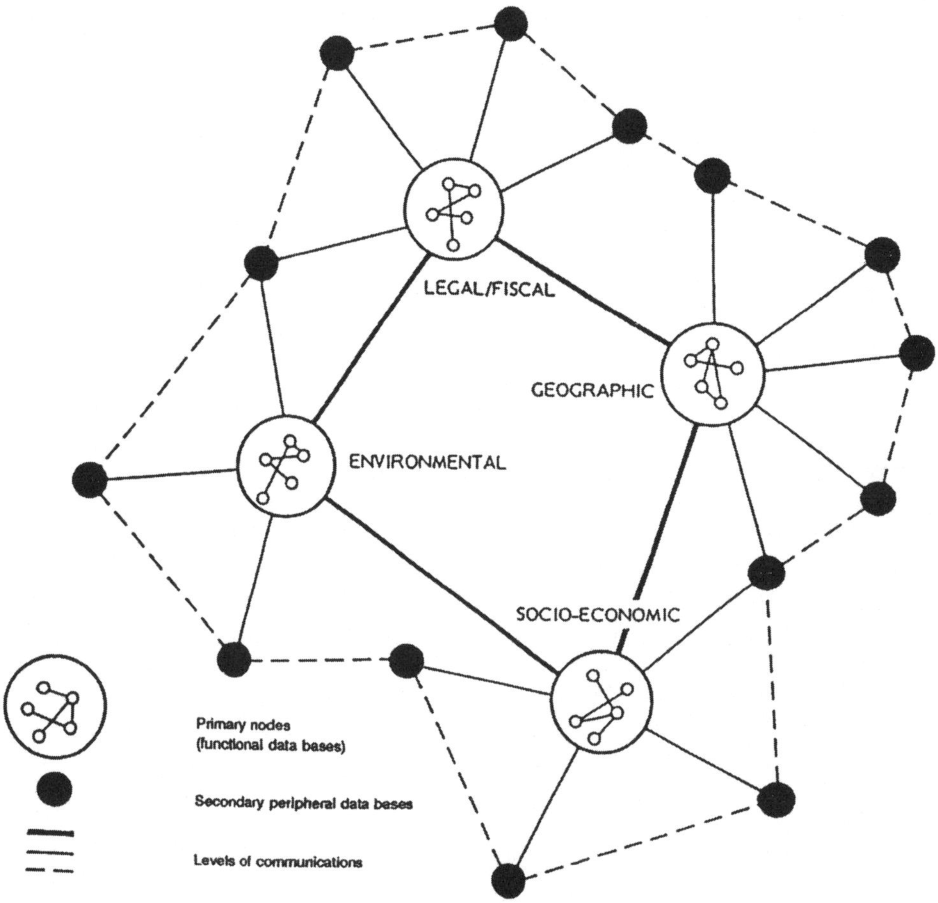

Figure 5.2 Structure of the South Australia land information system.
Source: Sedunary, 1987, p. 138.

In 1991 the state of Victoria decided that the development of multi-agency linked geographic information systems was a major government priority and established an Office of Geographic Data Coordination (OGDC) to coordinate geographic information management in the state (Chan and Williamson, 1995). The OGDC is housed in the Department of Treasury and Finance of the State Government. As part of its overall remit the OGDC produced a 10-year GIS data management strategy in 1993 with the following objectives:

- contribute to increased economic growth for the state.
- create a flourishing information industry.
- support greater competitiveness in global markets.
- results in internal efficiencies across government.
- provide a major capital asset for Victoria through an essential state data infrastructure.

(OGDC, 1995, p. 1)

To achieve these objectives a framework has been developed which consists of four main components:

- A focus on the business information needs of government in four key programme areas: land administration; human services; natural resources; infrastructure planning.
- Three year implementation plan and funding programme to achieve the essential systems and outputs.
- A financial model and pricing policy to achieve sustainable maintenance, a return on capital, industry stimulation.
- Administration and quality assurance, including data directories, custodians, access, standards, distribution, maintenance, technology and benefits monitoring. (OGDC, 1995, p. 1)

With this in mind the Office of Geographic Data Coordination has identified 61 high priority information products for the four key programme areas that draw upon 270 separate data sets and assessed the cost benefits of developing these products (Alexander, 1995, p. 506). Work has already begun on 27 of these projects which is likely to produce benefits of the order of A$150 million (US$200 million) over six years (Alexander, 1995, p. 513).

The Office of Geographic Data Coordination has also set up a separate agency (Geographic Data Victoria) in conjunction with Melbourne Water to maintain and upgrade the State Digital Map Base which is a core element of all state requirements. Geographic Data Victoria operates as a commercial agency to package and market cadastral, topographic and road centre-line data sets. From the standpoint of geographic information coordination it is important to note that the overall priorities set up by the OGDC are recognised as a key ingredient in the state's budgetary process which influences the budget proposals that are put forward by individual departments and agencies (Price Waterhouse, 1995, p. 45).

The Australian Surveying and Land Information Group (AUSLIG)

The Australian Surveying and Land Information Group was formed by the merger of the Australian Survey Office and the Division of National Mapping within the Commonwealth Department of Administrative Services in 1987. Its head office is in Canberra and there are also offices in the other state/territory capital cities and in Townsville, Queensland. As a business unit within the Department of Administrative Services AUSLIG provides a wide range of services for both public and private sector clients in the fields of mapping, surveying, geodesy and remote sensing.

AUSLIG is also responsible for national-level mapping activities and geographic information coordination. Its digital topographic data products include national coverage at the 1:250,000 scale and it began work in early 1996 on a 1:100,000 scale product. Its Public Interest Policy and Coordination section is the focus for geographic information coordination at both the (trans)national and state levels, providing the secretariat for both the Australia New Zealand Land Information Council (ANZLIC), whose activities will be discussed in some detail later in the chapter, and the Commonwealth Spatial Data Committee as well as the Intergovernmental Committee on Surveying and Mapping (Baker, 1995b).

In August 1996 the Commonwealth minister for administrative services announced that 'AUSLIG's existing commercial activities will be tested for sale through a management buyout or trade sale, tested concurrently. Remaining commercial activities and CSO funded activities will be market tested for outsourcing, with AUSLIG undertaking the expert buyer function' (Department of Administrative Services, 1996). However, it will continue to act as the Commonwealth focal point for geographic information coordination.

The Australian Bureau of Statistics (ABS)

The precursor of the Australian Bureau of Statistics, the Commonwealth Bureau of Census and Statistics, was set up in 1906. The present Bureau is the central statistical agency for the Commonwealth government and also provides statistical services for state and territory governments. Its mission is 'to assist and encourage informed decision making, research and discussion within governments and the community by providing a high quality, objective and responsive national statistical service' (McLennan, 1995, p. 4). The head of the Australian Bureau of Statistics is the Australian Statistician who is appointed by the Governor General.

ABS employs about 3,500 people of whom 1,500 work in its head office in Canberra. The remainder of its staff are employed in branches in the states and territories. In addition to their state-wide responsibilities, each of these branches acts as the primary centre within ABS for a particular set of statistics. For example, Tasmania is responsible for agricultural statistics.

Over the last few years ABS has developed a wide range of statistical products often in conjunction with private sector agencies. It was one of the pioneers of CD-ROM technology in Australia with its 1986 and 1991 census CDATA products and their Map Info and supermap GIS display capabilities. It is currently experimenting with the on-line delivery of basic statistical services to subscribing Australian university libraries via the Internet.

In the last few years it has extended its range of spatial data products considerably. One of the most interesting of these new products is the Integrated Regional Data Base (IRDB) which was launched in 1993 to make a wide range of data available in the form of spreadsheet-like tables at the Statistical Local Area level. Since then a new version of this database has been developed which includes a greater expanded and regionally integrated data range derived not only from ABS data but also data collected by other Commonwealth agencies. This version has also a greatly expanded functionality including geographical browsing, querying and mapping capabilities (Laughlin, 1994). A particularly interesting feature of this database is the procedures it includes for taking account of areas that have undergone boundary changes in relation to time series data.

ABS also provides a number of customised services. These include a variety of community profiles for any given area derived from its census data as well as its 4-site product which provides demographic and business data for customer-defined areas, together with a comparison between these data and data for any other selected area as well as a report on current trends in the economy at the national level.

A census of population and housing is carried out every five years in Australia by ABS. The most recent census took place in 1996. This broke new ground with respect to the digital map base that was created for census design, mapping and collection purposes. The task of creating an integrated national digital base of Australia at the level of detail required by ABS was given to a consortium of public sector mapping agencies (PSMA) in 1993. Under the terms of its contract PSMA was required to collect the necessary data from the state and territory agencies and supply it in a standardised form to the private company selected by ABS to develop a census mapping system for the 34,000 census collection districts in the country. The work of PSMA in connection with this project is described in greater detail in the section on core data later in this chapter.

INSTITUTIONAL CONTEXT

Disseminating Geographic Information

It has long been recognised in Australia that land information has considerable potential for income generation. However, most of the agencies involved are also aware that many of the normal economic considerations governing the dissemination of goods and services are not necessarily valid for information because the normal rules about scarcity do not apply to information. Notwithstanding this, 'increasingly there is a view that as a general principle "user pays" philosophy is appropriate for supply of land related information' (ALIC, 1990c, p. 3).

With these considerations in mind, the national council responsible for coordinating land information (ANZLIC) was given the task under the Intergovernmental Agreement on the Environment in 1992 of facilitating 'the co-ordination of intergovernmental arrangements (including appropriate financial arrangements) and [providing] mechanisms to make the data more accessible across all levels of government and the private sector' (CSDC, 1994, p. 6). This resulted in a draft national policy for the transfer of land-related data which was approved by ANZLIC in October 1993 and submitted to the prime minister in February 1994 with a view to getting the agreement of all first ministers. The basic principle underlying this policy is as follows:

> Government spatial data, already collected and funded in the public interest, could be generally made available at the average cost of transfer (i.e. the cost of distribution), subject to certain conditions. [However,] any collection, upgrade or further processing of public interest data to meet client needs may be subject to additional charges above and beyond recovering the full cost of distribution. (CSDC, 1994, p. 7)

This draft policy applies only to public interest usage, and data for commercial usage will be generally made available at rates in accordance with government directives on charging. In practice, however, pricing policies with respect to the use of digital cadastral databases by non-government agencies vary considerably from state to state and from territory to territory as can be seen from Table 5.2. This indicates that while some states, such as South Australia and Western Australia, set their prices low to encourage users,

Table 5.2 Comparison of data charges for non-government users in different states and territories

	Number of parcels (million)	Charge for non-government users (AUD/parcel)†
New South Wales	3.15	1.75
Queensland	1.2	2.0
South Australia	0.8	0.5
Tasmania	0.3	0.8
Victoria	2.3	0.91
Western Australia	0.86	0.3
Australian Capital Territory	0.1	0.9
Northern Territory	0.05	*

* No specific policy.

† Australian dollars. For a rough conversion to US dollars, multiply by 1.33.

Source: Wan and Williamson (1995a, p. 50)

others, such as New South Wales and Queensland, try to recover some of the additional costs of data capture and maintenance in their pricing (Wan and Williamson, 1995a, p. 50). As a result the average cost to non-government users in the latter may be as much as six times that of the former.

At another level there is growing pressure on many state and territory governments to think more strategically about their corporate geographic information resources. With this in mind, for example, the South Australian state government commissioned a study to evaluate the potential for a state information industry to contribute to regional economic development in June 1995 (Johnson, 1995). In Victoria the thinking behind the Office of Geographic Data Coordination has been summarised in the following terms:

> While few would attempt to construct a $10 million dollar building without detailing mandatory functional requirements such as purpose, shape, form and finish through the skills of an architect, many geographic information systems costing about the same order of magnitude evolve as small pilot systems that grow into fragmented, uncoordinated collections of geographic and textual databases. The outcome of investment in geographic information systems is far less visible than other forms of capital investment, but it should be no less accountable. (Alexander, 1995, p. 512)

The application of cost recovery principles is also reflected in the marketing strategies of the Australian Bureau of Statistics. Its charging policy is designed to meet three main objectives:

- to enable the demand for ABS products and services to be used as a more reliable indicator of how ABS resources should be used;
- to encourage users to address their real needs for ABS products, both statistics and services; and
- to relieve the general tax payer of those elements of the costs of the statistical service which have a specific and identifiable value to particular users. (McLennan, 1995, p. 19)

Consequently, ABS pricing policy balances public good obligations against 'user pays' criteria. The costs of producing the public good copies of ABS publications is funded through the budget appropriation of ABS. However, users requiring additional information must pay the full costs beyond the costs of collection and the production of clean unit files from which those data are derived. Furthermore, to meet the needs of the market for statistical information:

> The ABS needs to invest resources in the development, production and delivery of such products and services, and the price of such products and services are set at market prices where quantifiable, but at least to recover the full costs involved, including amortisation of capital and a reasonable allowance for contingencies and risk involved.
> (McLennan, 1995, p. 20)

Legal Protection

Generally developments in copyright law in Australia, as elsewhere, lag behind recent technological developments in connection with the storage, retrieval and dissemination of computer-held databases (ALIC, 1990b, p. 19). The Australian Copyright Act of 1968 (as extended by the Copyright Amendment Acts of 1980 and 1984) protects authors' rights in geographic information products. In essence, 'Copyright can subsist in an original land information database where skill, judgement and labour has been expended in the representation of data in databases' (ALIC, 1990b, p. 7). In the case of geographic information, copyright in databases acquired directly by government agencies rests with the Crown. However, the ownership of copyright in databases containing data extracted substantially from documents lodged with these agencies (e.g. land titles) is open to question.

Under these circumstances it is recommended that government agencies holding geographic information should release it only under a licence to use and not transfer ownership. 'Agencies should license the use of land information related computer programmes and computer based data in their custody to clients under an established form of agreement that specifically defines the uses of the data licensed to the clients' (ALIC, 1990b, p. 12).

The 1988 Australian Privacy Act set standards for information collected and held by Commonwealth agencies. In addition some states have introduced legislation on information and privacy matters. Much of this legislation is designed to protect the privacy of the individual or business whereas the emphasis of geographic information systems is on land-related data. Nevertheless, there is a fine boundary between land-related data and personal data. Consequently there is a need for custodians of such databases to take special steps to protect both individuals and groups from the abuse of information held under their control (ANZLIC, 1992b, p. 7).

ELEMENTS OF NATIONAL GEOGRAPHIC INFORMATION STRATEGY

Coordination

The Australian Land Information Council (ALIC) was established in January 1986 by agreement between the Australian prime minister and the heads of the state governments to coordinate the collection and transfer of land-related information between the different levels of government and to promote the use of that information in decision-making (ANZLIC, 1992a, p. 1). In November 1991 New Zealand became a full member of the Council which was renamed the Australia New Zealand Land Information Council (ANZLIC).

Each of the 10 members of ANZLIC represents a coordinating body within their jurisdiction (i.e. the Commonwealth Spatial Data Committee (CSDC), the relevant coordination bodies at the state and territory levels and Land Information New Zealand (LINZ)). These members have the responsibility for both expressing that jurisdiction's view at the Council and also promoting ANZLIC's activities within their jurisdiction. The Council also maintains close links with other relevant coordinating bodies such as the Intergovernmental Committee on Surveying and Mapping (ICSM) and industrial bodies such as the Australasian Urban and Regional Information Systems Association (AURISA).

The responsibilities of ANZLIC for national coordination are not restricted to cadastral matters but cover all types of land information including socio-economic data, natural resource information, environmental data and utilities and infrastructure information (ANZLIC, 1992a, p. vii). During its lifetime the Council has produced a number of major reports on the status of land information in Australia (see, for example, ANZLIC, 1992a) and the series of position papers on the institutional context of land information in Australia discussed in the previous section, as well as three versions of their national strategy for the management of land and geographical information in 1988, 1990 and 1994.

ANZLIC is serviced by the Public Interest Policy and Coordination section of the Australian Surveying and Land Information Group (AUSLIG) which is part of the

Department of Administrative Services within the Commonwealth government in Canberra (see p. 60). The most recent of the three national strategy documents prepared by ANZLIC (1994) is its strategic plan for the period 1994–7. The basic vision underlying this strategic plan is as follows:

> Australia and New Zealand will have the land and geographic data infrastructure needed to support their economic growth, and their social and environmental interests, backed by national standards, guidelines, and policies on community access to that data.
>
> (ANZLIC, 1994, p. 5)

With this in mind, five key objectives are defined:

- Data infrastructure
  To provide the fundamental land and geographic information infrastructure needed to support the economic growth, and social and environmental interests of Australia and New Zealand.
- Standards
  To provide the national standards and guidelines necessary to enable the effective use and integration of land and geographic information.
- Access
  To maximise community access to land and geographic information with due regard for issues of privacy and confidentiality.
- Industry development
  To support development of the Australia and New Zealand land and geographic information industry.
- Organisational framework
  To strengthen the land and geographic information organisational framework.

(ANZLIC, 1994, p. 5)

With respect to the data infrastructure itself, the primary task is seen as defining the fundamental or core data sets, implementing an infrastructure containing these data sets and establishing a reporting and revision service to respond to changing data needs. Insofar as standards are concerned, ANZLIC will lead the development and implementation of national standards and guidelines for land and geographic information, including standards for data transfer, data quality, data directories and data models through its jurisdiction coordination committees (ANZLIC, 1994, pp. 8–10).

To maximise access to geographic information ANZLIC plans to develop a national geographic metadata directory system (see below) as well as take steps to minimise barriers to community access to geographic information, with due regard for issues of privacy and confidentiality. In terms of industry development ANZLIC intends to identify the economic benefits of building a geographic information infrastructure as well as identifying education and training needs, encouraging a continuing R&D programme and promoting international cooperation (ANZLIC, 1994, pp. 11–13).

To strengthen the organisational framework ANZLIC is taking steps to raise awareness of its role as the peak council for geographic information in Australia and New Zealand and using its best endeavours to ensure that political support for its activities is obtained and maintained while developing closer links with local government at both the national and jurisdiction levels.

The key components of the national land and geographic data infrastructure have been spelt out in greater detail by the secretary of ANZLIC, Graham Baker, with respect to what he identifies as its four key components:

- [The] institutional framework [which] defines the policy and administrative arrangements for building, maintaining, accessing and applying the standards and data sets.
- Technical standards [which] define the technical characteristics of the fundamental data sets.
- Fundamental data sets [which] are produced within the institutional framework and fully comply with the technical standards.
- [The] distribution network [which] is the means by which the fundamental data sets are made accessible to the community, in accordance with policy determined within the institutional framework, and to the technical standards agreed. (Baker, 1995a, p. 4)

The institutional framework for the data infrastructure can be further subdivided into a number of key elements. These include the institutional structure to lead the development of the national geographic information infrastructure. This is essentially the task of ANZLIC and its associated committees. Another key element is the notion of custodianship which defines the responsibilities and rights of the lead agencies who will be charged with developing the specifications of the data sets themselves in consultation with the users (ALIC, 1990a).

A great deal of work has already been undertaken in Australia on the definition of technical standards relating to the geodetic datum, data models, data dictionaries, data quality, data transfer and metadata. ANZLIC has also supported the work of Standards Australia and Standards New Zealand in the belief that compliance with a single set of integrated national standards will be more efficient for industry than having to comply with different standards developed by different agencies. The council is also actively involved in the work of ISO TC/211 at the international level. It is also worth noting that considerable progress has already been made in adapting the US Spatial Data Transfer Standard (SDTS) to Australian conditions in the AS/NZS4270 which was published in January 1995 (Baker, 1995a, p. 10).

In the view of ANZLIC a fundamental data set is one 'that is critical to the objectives of more than one agency' (Baker, 1995a, p. 10). With this in mind, ANZLIC is developing a methodology to determine corporate government needs and priorities in relation to perceived economic, social and environmental benefits. The basic objective of this methodology is to identify multiple high-benefit information products. Data sets that will be considered in this exercise are likely to include 'aerial and satellite imagery, the Cadastre, census results, land use and land cover, place names, administrative areas, transportation networks, utility networks, coastline, rivers and lakes, elevation, soils, vegetation, geology, climate, pollution, hazardous sites, and areas of environmental significance' (Baker, 1995a, p. 11). With this in mind, it is argued that issues of custodianship and the funding of production and maintenance programmes can only be addressed when this exercise is complete.

A task that is likely to be of particular importance under the general heading of the distribution network is the specification of a technical framework to give the community access to fundamental data held upon a number of independently maintained systems. This will require the use of a wide range of communications technology devices to deliver these services.

In essence, then, ANZLIC's view of national geographic information strategy can be summarised as follows: 'The primary objective of a national data infrastructure is to ensure that users of land and geographic data who require a national coverage, will be able to acquire complete and consistent data sets meeting their requirements, even though data is collected and maintained by different jurisdictions' (Smith and Thomas, 1996, p. 5).

Core data

As noted above, considerable progress has been made in connection with the creation of core databases based on the digital cadastral databases and the digital topographic databases held by each of the eight state and territory governments. Table 5.1 shows that all the states and territories have completed their digital cadastral databases with the exception of Tasmania which plans to finish this stage in 1998. The examples from South Australia and Victoria also show that these databases typically contain not only land titles and map information but also environmental and socio-economic data. Figure 5.3 contains an extract from the digital cadastral database prepared by Geographic Data Victoria. Sovereign Hill was a goldmine one hundred years ago and is the state's most famous tourist attraction.

Given the development of different core databases in the eight Australian states and territories, the inevitable question that arises is to what extent they collectively function as a core national topographic database at the present time? A partial answer to this question can be found in the measures that were taken to establish a Commonwealth-wide digital map base for the Australian Bureau of Statistics 1996 Census of Population and Housing.

As noted above, a consortium of public sector mapping agencies (PSMA) was set up in 1993 to carry out this task. The findings of this project demonstrate that, while each jurisdiction has recognised the need for greater coordination of their own data holdings, little progress had actually been made in integrating these data to create a national spatial data infrastructure (Mooney and Grant, 1995, p. 3). However, despite the lack of nationally agreed standards, interim standards were put together in a matter of weeks and the integrated data set was delivered to ABS on time in October 1995. Mobbs (1996, p. 4) estimates that the task of integration took 26 months with a team of six to seven operators at the Land Information Centre at Bathurst in New South Wales alone, representing approximately 15 person years' work. The data set contains 1.9 gigabytes of topographic data in 1,836 tiles and 2 gigabytes of cadastral data in 2,407 tiles at source scales that vary from 1:500 for densely built-up urban areas to 1:250,000 for remote regions (Mobbs, 1996, p. 5).

PSMA's mission statement is 'to return economic benefits to the Nation through the preparation and exploitation of a national topographic map database'. Following the completion of the 1996 census project ANZLIC has approved two motions expanding the work of PSMA with respect to the creation and maintenance of this core database. These require the following:

- That the PSMA expand its role to make the PSMA dataset available to users under conditions to be determined by the Board of Management.
- That the PSMA has a role in establishing the mechanisms for enhancing the database with other appropriate national datasets. (Mobbs, 1996, p. 5)

Metadata

One of the components of the current ANZLIC strategic plan is to build upon previous efforts at the state and Commonwealth levels to develop a national land and geographic data directory system for Australia and New Zealand. These include projects such as the Queensland Land Information Directory (QLID) which contains 25,000 entries for data

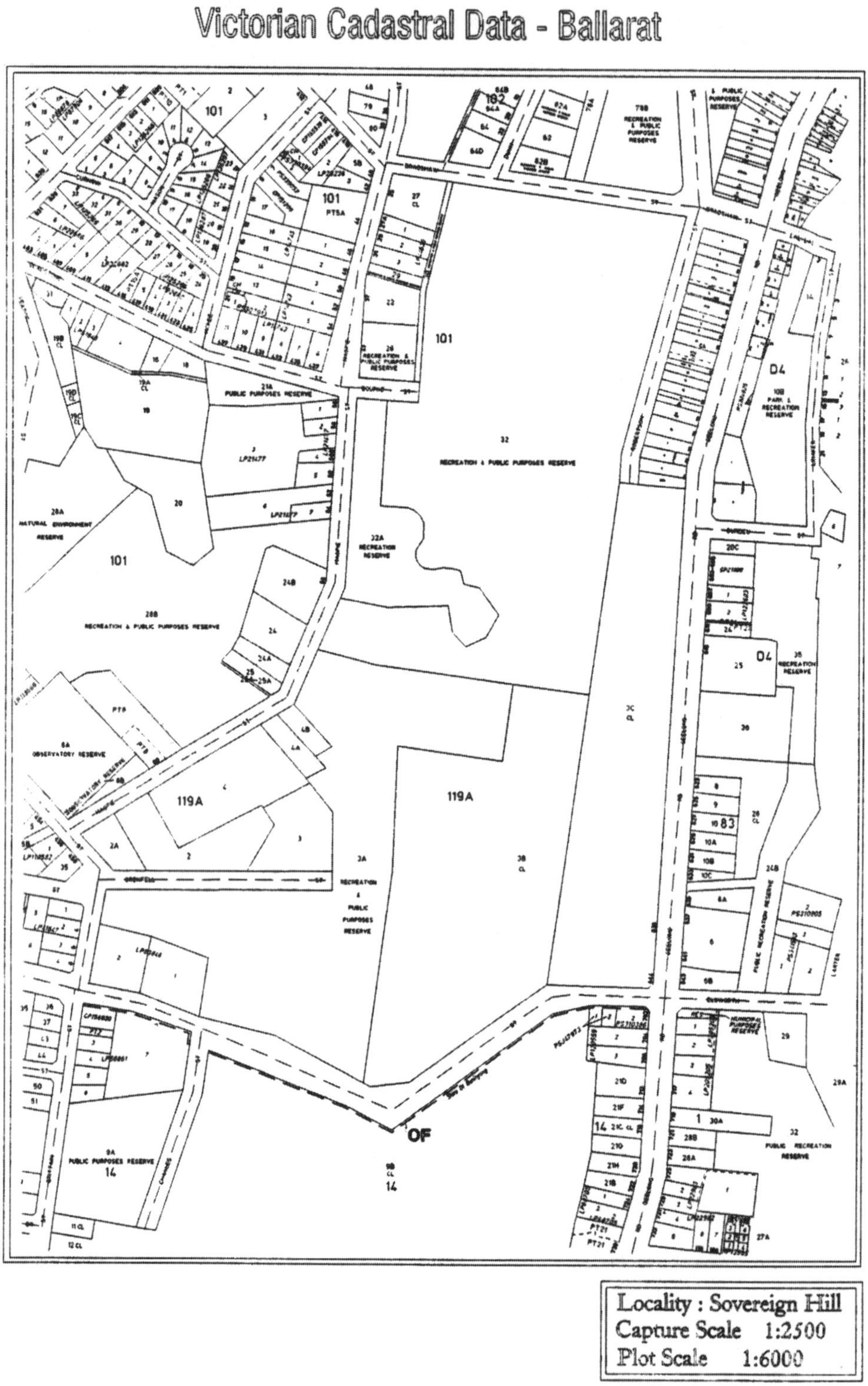

Figure 5.3 Example of State of Victoria cadastral database for Sovereign Hill in Ballarat. Crown (State of Victoria) copyright. Reproduced with permission of Geographic Data Victoria, Department of Natural Resources and Environment.

products from 46 agencies (Abel, 1996) as well as national projects such as the on-line National Directory of Australian Resources (NDAR) and the Environmental Resources Information Network (ERIN) set up as a central resource within the Department of Environment, Sport and Territories. With this in mind, the following policy on the transfer of metadata was adopted by ANZLIC in 1994:

- Jurisdictions will contribute metadata to the national directory at no cost;
- Core metadata will be made available from the national directory to contributing jurisdictions at no cost;
- Mechanisms should be established by jurisdictions to allow any potential user to access metadata freely and readily. This does not preclude the development of add on services on a fee-for-service basis. (ANZLIC, 1996, p. 2)

ANZLIC has adopted a 'pages' concept for a national metadata framework where more general information is recorded at the highest level and additional information at lower levels within a hierarchical structure. Consequently only the highest level information need be held by Commonwealth and state governments and more detailed information can remain in the hands of the data custodians.

To make such a system work, the completion of a limited number of core metadata elements for each higher-level data set will be mandatory. These will be consistent with the corresponding elements of the Content Standards developed by the Federal Geographic Data Committee in the United States to promote international compatibility and also to take advantage of software tools developed by US-based GIS vendors for this purpose. Core metadata elements suggested by the ANZLIC group include the following:

- Dataset: i.e. title, custodian and jurisdiction
- Description: i.e. abstract, search words, geographic extent
- Data currency: i.e. beginning and end date
- Dataset status: i.e. progress, maintenance and update frequency
- Access: i.e. stored and available formats, constraints on access (if any)
- Data quality: i.e. lineage, positional and attribute accuracy, logical consistency and completeness
- Contact information. (ANZLIC, 1996, p. 5)

EVALUATION

The findings of this analysis highlight the extent to which a Commonwealth dimension has come into being to coordinate the geographic information strategies that have been evolving over the last two decades at the state and territory levels. As a result of the land administration system that has developed in Australia at the state and territory levels responsibility for geographic information of all kinds is highly centralised and there is a tradition of cooperation between the different public agencies involved. Given the federal system, there is also a tradition of interprofessional cooperation at the Commonwealth level. Nevertheless, Australia has eight different cadastral systems and eight different surveying and mapping systems in operation at the state and territory level at the present time.

The establishment of a consortium of public sector mapping agencies (PSMA) to create a national digital database for the 1996 Census of Population and Housing must be regarded as a significant step towards an integrated core database for Australia as a whole. As the motion approved by ANZLIC indicates, the next steps for PSMA are to build upon the experience built up during the 1996 census project and to expand its role to make this data set more widely available.

The brief descriptions of the activities of the two state-level land information coordination agencies indicate the contrasting positions that have been taken up in the different states and territories regarding the coordination of geographic information-related activities. Whereas the model developed by Victoria can be regarded as a focused mechanism for prioritising funding proposals, which is an important input to the state budgetary process, it has been argued that in some other states and territories existing coordination arrangements are *ad hoc* or incomplete and in need of reform (Price Waterhouse, 1995, pp. 44–5).

The work of ANZLIC as the national coordinating body for geographic information builds upon these traditions. Its plans clearly spell out the key elements of national geographic information strategy for Australia and the Council has been very active in filling in the detail with respect to the development of national standards and more recently the framework for metadata services. It has also played a vital role in raising overall awareness of the institutional issues involved in disseminating geographic information and the need to protect the legal rights and responsibilities of data custodians. However, it must also be recognised that ANZLIC relies on the achievement of consensus to realise its objectives and that it has no executive powers to compel its members to comply with its policies. The inherent weakness of its position is reflected in the draft National Policy for the Transfer of Land Related Data following an extensive process of consultation including all ANZLIC members, the consultative structures that they separately represent and many other national coordinating bodies with an interest in geographic information. It has been argued, therefore, that such a policy represents 'the lowest common denominator' on which consensus can be found (CSDC, 1994, p. 6).

Despite the recognition of the public good dimension of the core cadastral and topographic databases created by the state and territory administrations, there are considerable pressures on all public sector agencies in Australia to cover some or all of the costs of their non-public interest activities. Such activities are likely not only to limit access to some of the land and statistical information held by public agencies to those who are able to pay for the services but also to raise questions about the need to establish access rights and rules in the context of national competition policy. It can be argued that many of the general principles put forward in the Hilmer review of national competition policy (Hilmer, 1993) for the reform of the natural monopoly element of telecommunications and energy supply networks also apply to the provision of core geographic data sets. For example, the need to ensure transparency of the application of competitive neutrality principles was a key factor in the decision to restructure AUSLIG. More generally, base/primitive data compiled by agencies as part of their core function of government may also need attention because of the following:

- their natural monopoly characteristics;
- the strategic position of suppliers, in as much as access to the data is required by downstream users wishing to compete in the market for value added products; and
- those suppliers competing or being in potential position to compete in the provision of value added products and therefore having an incentive to inhibit access by competitors.

(Price Waterhouse, 1995, p. 53)

It should also be noted that the ownership of copyright in many databases of this kind which contain information derived substantially from documents lodged with public agencies is open to question.

The findings of the analysis also show that Australian users have become very sophisticated and are demanding more from the DCDBs. A recent survey of key users in the utilities and local government by Wan and Williamson (1995a) identified deficiencies in the areas of data quality, data structure, data access, data transfer and data maintenance. At the same time the design model used for some states has been criticised by Barnes (1995) on the grounds that the central position occupied by land ownership as the hub activity inhibits the development of environmental and socio-economic applications and also creates tensions between the surveying and the project user professions.

CHAPTER SIX

The United States

The National Spatial Data Infrastructure in Perspective

INTRODUCTION

The United States is one of the largest countries in the world with a population of more than 250 million living on a land area of over 9,350,000 square kilometres stretching from the Atlantic Ocean on its east coast to the Pacific Ocean on its west coast (see Figure 6.1). The 13 original states gained their independence from Britain in 1783 and the number of member states has risen to 50 in 1959 when Alaska and Hawaii joined the Union. The history of the mainland United States closely reflects its settlement which began on the east coast and spread across the nation during the 19th century, with the last two members of the contiguous mainland group of states, Arizona and New Mexico, joining the Union as recently as 1912.

One of the most distinctive features of the United States is the large number of agencies involved in creating geographic information. As might be expected, given the federal structure of the US government, many important responsibilities for geographic information are dealt with at the state and local government level and there are wide variations between states in the way that these responsibilities are carried out. Particularly important from this standpoint are land title registration and land taxation matters which rest with county and local level governments in each state. As a result, as many as 80,000 agencies of this kind are involved in some way with geographic information creation.

At the federal level many different agencies are involved in mapping and geographic information collection activities. Although the US Geological Survey National Mapping Division is primarily involved in the creation of the small-scale national digital topographic database, the National Imagery and Mapping Agency is also heavily involved in mapping activities. It should also be borne in mind that some federal agencies manage substantial holdings of land, particularly in the western parts of the United States. Overall, about one-third of all land is owned by federal agencies and in states such as Nevada this proportion is over nine-tenths. The most important agencies in this respect are the Department of the Interior Bureau of Land Management, which administers over a million square kilometres of land in federal ownership, and the US Forest Service, which manages a further three-quarters of a million square kilometres of the surface of the United States (National Research Council, 1993, pp. 31–43).

Figure 6.1 The United States.

With these considerations in mind, the next section considers state and county level land information activities in the United States together with the small-scale mapping tasks carried out by the US Geological Survey National Mapping Division and the work of the Bureau of the Census. The organisations that have emerged to coordinate these key players at the federal level will be dealt with later in the section on national geographical information strategies.

MAIN PROVIDERS OF GEOGRAPHIC INFORMATION

State and Local Government Land Administrations

As noted above, land titles records and land taxation responsibilities are devolved in the United States to local government and the way in which these are discharged varies considerably from state to state and even between different areas of the same state. According to Huxhold (1991, p. 187), as far as land titles are concerned:

> The United States has an elaborate system to help ensure that legal claims of interest in property are either transferred or satisfied when the ownership of the property changes. This system is generally referred to as a land title recordation system and is based on the official recording of such documents by local government (usually a County Recorder or Register of Deeds or a City or Town Clerk with similar responsibilities).

Similarly the appraisal of property for taxation purposes and the collection of taxes on real estate are performed either by the county, the township or the municipality depending on the state or the region within the state.

Under these circumstances it is not surprising that the desire to coordinate these activities, coupled with the need to provide both technical and financial assistance to local government agencies, has featured prominently on the agendas of state legislatures for the last 20 years, and the experience built up in this way is reflected in the evolution of the national geographic information strategies that have emerged at the federal level.

The geographic information activities of each of the 50 states have been reviewed in the compendium compiled by Warnecke (1992) for the Council of State Governments. This highlights the diversity of experiences and the variations in practice that reflect in particular the different histories of the original group of states that joined the Union and those of the 30 western states whose initial settlement was controlled to a large extent by the requirements imposed by federal government through its Public Land Survey System (PLSS) administered by the Bureau of Land Management.

Under these circumstances it is impossible to select any one state as a typical example of state-wide and local government land information activities. Notwithstanding this, the experience of the state of North Carolina illustrates some of the main features of such activities. Like many other states it has a long tradition of land records mapping even though it lies outside the PLSS regime. North Carolina also has extensive state-wide geographic information coordination activities and there is strong political support at the state level for such activities. However, it is also unusual in the extent to which state-wide coordination activities are cross-subsidised by revenue generation through the provision of customised products and services.

North Carolina was one of the 13 original states of the Union. It has a land area of 137,000 square kilometres and a population of over 6.5 million. North Carolina is the leading industrial state of the south Atlantic states and is becoming increasingly high tech in character with a concentration of activities at Research Triangle Park near the state capital, Raleigh, and at the University Research Park near Charlotte. Despite this, more than half its population live outside the urban areas, giving it one of the largest rural populations in the nation.

North Carolina has a long history of state-wide geographic information activities beginning in 1974 when the state legislature passed a Land Policy Act which set up a Land Resources Information Service (LRIS) to deal with the 'lack of systematic collection, classification, and utilisation of information regarding the land resource' (Warnecke, 1992, p. 309). In 1977 a further statute set up a Land Resources Management Programme to provide technical and financial assistance to local governments to enable them to modernise their land records. Since that time the state has provided assistance to local officials with respect to a wide range of activities including base mapping, cadastral mapping, standardisation of parcel identifiers and the automation of land records. This programme has resulted in the completion of land parcel mapping in 82 out of the 100 counties in the state and most of these have subsequently converted these data into digital form for use in GIS applications (North Carolina Geographic Information Coordinating Council, 1994, pp. 1–6).

A good example of these efforts is the land parcel database maintained by Wake County which reflects standards guidance from the State Land Records Office. Wake County covers 12 incorporated municipalities in the greater Raleigh area with a combined population of over half a million. The conversion of its land parcels database to digital format was completed in 1991 and the database currently contains details of over 200,000 parcels mainly on 1:1,000 scale maps. Wake County is one of the fastest growing areas in the state and about 5,000 new parcels are added to the database each year as a result of subdivisions.

Under an Executive Order signed by the governor on 30 July 1991 the responsibilities of the Land Resources Information Service were incorporated into a state-wide Centre for Geographic Information and Analysis in the Governor's Office of State Planning. At the same time a Geographic Information Coordinating Council was established consisting of 12 members initially drawn from the main GIS user departments at state level.

Subsequently, a further Executive Order dated 21 May 1994 expanded the range of membership of the Council to enable a wider representation from local and regional government agencies in the state. In its current strategic plan the Council (*ibid.*, pp.1–3) identified three main outcomes from state-wide coordination activities:

- A corporate geographic database which will be distributed, used and maintained by multiple government agencies and other organisations throughout the state.
- A comprehensive electronic directory describing data held in the corporate geographic database and other geographic information held by multiple government agencies and other organisations throughout the state.
- A GIS partnership programme which will forge a cohesive system of relationships among users of geographic data throughout North Carolina.

In mid-1996 the corporate geographic database which is administered by the Centre for Geographic Information and Analysis contained about 65 separate layers of geographic information relating to the state. Two-thirds of these are held at the Centre itself and the remainder are available from the original custodians over a distributed network. The Centre has also become a node in the National Geospatial Data Clearinghouse that is being developed as part of the National Spatial Data Infrastructure (see p. 87) and is one of the 24 partnership case studies for the National Digital Geospatial Data Framework (see p. 86).

As a result of its increased coordination activities the number of staff in the Centre has increased to 35 in recent years. However, the Centre receives no direct funding from the state for these activities and operates essentially on a cost recovery basis through the provision of customised services and the development of geographic information products for a wide range of state, federal and local government agencies. It should be noted, however, that cost recovery principles are not applied to the dissemination of material from its corporate geographic database which is governed by a North Carolina state public records law ruling that such information must be made available at no more than the costs of dissemination.

United States Geological Survey National Mapping Division

The United States Geological Survey (USGS) is part of the Department of the Interior which has overall responsibility for the management of nationally owned public lands and natural and cultural resources. The USGS was set up on 3 March 1879 charged with the 'classification of the public lands, and the examination of the geological structure, mineral resources, and products of the national domain' (US Geological Survey, 1994, p. 9). Its current work is organised around its geological, national mapping and water resources divisions.

The National Mapping Division (NMD) is the smallest of the three divisions. At the start of 1996 it employed around 1,500 staff at its headquarters in Reston, Virginia, and in its four regional mapping centres in Reston; Rolla, Missouri; Denver, Colorado, and Menlo Park in California. Its budget appropriation from the federal government in 1995 was $124 million. Its national mapping programme strongly reflects the interests of the Department of the Interior as a whole and the USGS in particular. As the Mapping Sciences Committee of the National Research Council (1990, p. 8) pointed out in their review of the work of the NMD, 'for many decades the mapping it carried out was perceived as a support service for determining the geological and hydrologic nature of the country'. Despite this, however, the NMD houses the secretariat of the Federal Geographic Data Committee which is implementing the National Spatial Data Infrastructure (see p. 84).

The National Mapping Division produces a series of topographic map products ranging from the 1:24,000 to the 1:2,000,000 scales. Many of these products are created as a result of co-ventures between the NMD and other groups in the Department of the Interior, other federal agencies and state and local governments. Its flagship product is the 1:24,000 series of 54,000 topographic maps covering the United States as a whole which was completed in 1990. These maps are commonly known as 7.5 minute quadrangle maps because each map covers a four-sided area of 7.5 minutes of latitude and 7.5 minutes of longitude.

Work is now under way to develop a national digital cartographic database. This consists of digital line graphs (DLGs) of nine key features and digital elevation models (DEMs) of topography. By the end of September 1995, over 70,000 1:24,000 DLGs had been archived into the national cartographic database. These included 10,532 hydrographic overlays, 10,433 transportation overlays, 19,871 boundaries overlays and 17,149 Public Land Survey System overlays. Similar activities are under way in connection with the creation of DLGs for the 1:100,000 and 1:2,000,000 scale maps. In the case of the former the transportation and hydrography layers are already available in digital form as a result of the agreement reached with the Bureau of the Census in connection with the 1990 Census of Population (see below).

In addition to these activities the NMD has recently introduced two new digital products. A series of more than 200,000 Digital Orthophoto Quadrangles (DOQs) covering the country as a whole at the 1:12,000 scale is being produced as part of the National Digital Orthophoto Programme which is administered by the NMD together with the Natural Resources Conservation Service, the Consolidated Farm Services Agency, the Forest Service and the National States Geographic Information Council. So far 20,000 DOQs have been archived in the National Digital Cartographic Database and 30,000 more are in the process of production (see Figure 6.2). At the same time a digital raster product is being created by scanning the 1:24,000 analogue maps. It is anticipated that over half the country will be covered by this product by the end of 1996.

Most of these products are available at a nominal price which covers only the costs of dissemination and they are also free from copyright restrictions. Some selected USGS products have already been made available over the Internet at no charge. These include the 1:100,000 scale transportation and hydrography layers developed for the 1990 census.

At the present time the National Mapping Division is under considerable financial pressure which is making it necessary to reconsider its activities. The size of its appropriation from the federal government has declined in real terms over the last few years and its capacity to generate additional revenue through the sale of products and services is limited as most of its potential customers are federal agencies which also face similar pressures. Furthermore, the Division is also being required to contract out an increasing proportion of its mapping activities to private sector companies. It is anticipated that this will rise from 24 percent in 1995 to 60 percent by the end of 1998.

United States Bureau of the Census

The concept of regular census taking in the United States dates back to the framing of the original Constitution, which states in Article 1, Section 2, that political representation

Figure 6.2 Digital Orthophoto image of the Capitol Building in Washington DC. Courtesy USGS

in the House of Representatives will be apportioned every 10 years according to a population census. Consequently the census has been taken every 10 years since 1790 when Secretary of State Thomas Jefferson supervised the first census. The present Census Bureau dates back to 1902. It is housed in the Department of Commerce and regards itself as the fact finder for the nation.

The Bureau of the Census is a large and complex organisation whose activities are divided between demographic and economic statistics. It employs nearly 5,000 staff and its budget request for the 1997 financial year was of the order of $400 million. Over the last 30 years, the Bureau of the Census has developed into a primary user of both geographic and attribute data. It has pioneered large-scale mapping in cooperation with local governments and private companies by means of the development of the GBF/DIME (Geographic Base File/Dual Independent Map Encoding) files for the 1970 and 1980 censuses and the TIGER (Topologically Integrated Geographic Encoding and Referencing) files for the 1990 census. The development of the GBF/DIME files for the final stages of the 1970 census and their large-scale use in 1980 represented major steps towards the full automation of the Census Bureau's geographic support programmes.

An important distinction must be made between the GBF/DIME files and the TIGER data used for the 1990 census. Whereas the former covered only a small proportion of the national land area, the latter covered the whole nation as well as being grounded in rigorous conceptual models of topology and space (Marx, 1990). Unlike previous efforts, the creation of the TIGER database was the outcome of a landmark agreement with the US Geological Survey National Mapping Division in 1983 whereby the NMD provided the Census Bureau with scanned versions of the 1:100,000 scale maps for the 48 mainland states and the District of Columbia. In return the Census Bureau assigned cartographic classification codes and names for roads and other features in these computer files. This resulted in a win–win situation for both agencies: 'each agency would be able to accelerate its map production programme and, in the process, they would develop the first large scale digital map file of the US' (Sperling, 1995, p. 383).

It has been estimated that the cost of TIGER was of the order of $300 million (Blakemore and Singh, 1992, p. 31). However, it can be argued that this was justified not only in terms of the benefits obtained by both the Census Bureau and the USGS in relation to the execution of their responsibilities, but also in the stimulus that it provided for the development of the geographic information services market as a result of the residual cost/copyright free dissemination policies adopted by both the Census Bureau and USGS. This can be seen from the TIGER resource list which lists the names of 159 vendors who have notified the US Census Bureau of the capacity to process TIGER/files (US Geological Survey, 1996).

Plans for a similar exercise in connection with the 2000 census are now in hand (Bureau of the Census, 1996). The Census Bureau is continuing its partnerships with the US Postal Service and the US Geological Survey National Mapping Division to update its address and map information to support census data collection, processing and dissemination activities.

INSTITUTIONAL CONTEXT

Disseminating Geographic Information

In principle all information collected by federal government agencies in the United States must be made available to the public at a charge that covers only the costs of dissemination and with few restrictions under the provisions of the Freedom of Information Act (FOIA) and related legislation. Office of Management and Budget Circular A-130 on the Management of Federal Information Resources (OMB, 1996) provides a definitive statement of the position with respect to the dissemination of information collected by federal agencies. For this reason it is worth quoting at some length to highlight the issues involved:

> Section 7: Basic Considerations and Assumptions.
>
> (a) The federal government is the largest single producer, collector, consumer, and disseminator of information in the United States. Because of the extent of the government's information activities, and the dependence of those activities on public cooperation, the management of federal information resources is an issue of continuing importance to all federal agencies, State and local governments and the public.
>
> (b) Government information is a valuable national resource. It provides citizens with knowledge of the government, society, and economy – past, present, and future. It is a means to ensure the accountability of government, to manage the government's operations, to maintain the healthy performance of the economy, and is itself a commodity in the market place.

(c) The free flow of information between the government and the public is essential to a democratic society. It is also essential that the government minimise the federal paperwork burden on the public, minimise the cost of its information activities and maximise the usefulness of government information.

(d) In order to minimise the cost and maximise the usefulness of government information activities, the expected public and private benefits derived from government information should exceed the public and private costs of the information recognising that the benefits to be derived from government information may not always be quantifiable.

(e) The nation can benefit from government information disseminated both by federal agencies and by diverse non federal parties, including state and local government agencies, educational and other not for profit institutions and for profit organisations.

(f) Because the public disclosure of government information is essential to the operation of a democracy, the management of Federal information resources should protect the public's right of access to government information.

(g) The individual's right to privacy must be protected in federal government information activities involving personal information . . .

These general principles are also enshrined in open records laws at the state level in many states including North Carolina (see pp. 75–6). However, in practice, there is a growing tendency to alter the status of current open records laws at the state level to restrict public access and make it possible to sell geographic information products and services. The findings of surveys carried out by Dansby (1992) and Dando (1993) show that a number of states have already enacted some form of legislation of this kind. For example, the Iowa code of 1989, subsection 3, states:

> a government body which maintains a geographic computer database is not required to permit access to or use of the database by any person except upon terms and conditions acceptable the governing body. The governing body should establish rates and procedures for the retrieval of specified records, which are not confidential records, stored in the database upon the request of any person. (quoted in Dansby, 1992, p. 11)

In Kentucky, state law on the use of databases sets out criteria for cost recovery:

> Use of database or geographic information system.
>
> 1. A person who requests a copy of all or any part of a database or geographic information system, in any form for a commercial purpose shall provide a certified statement stating the commercial purpose for which it shall be used.
> 2. Such person shall enter into a contract with the owner of the database or the geographic information system. The contract shall permit use of the database or the geographic information system for the stated commercial purpose for a specified fee. The fee shall be based on the:
>   (a) cost to the public agency of time, equipment, and personnel in the production of the database or the geographic information system;
>   (b) cost to the public agency of the creation, purchase or other acquisition of the database or the geographic information system; and
>   (c) value of the commercial purpose for which the database or geographic information system is to be used. (quoted in Dansby, 1992, p. 11)

At the local level a large number of government agencies have developed geographic information marketing strategies whose pricing philosophies vary from straight cost recovery to market demand. According to Bryan (1993, p. 76), the best example of cost recovery is Knoxville/Knox County in Tennessee whose GIS data product pricing is

Table 6.1 Open access and revenue generation: some survey findings

	Clearly open access site*	Clearly revenue generation site*	Mixed or conflicting policy site*	No.	TOTAL %
National survey					
Cities	11	5	15	31	
Countries	8	8	27	43	
Multi districts	0	3	2	5	
Total	19	16	44	79	
Percentage	24	20	56		100
Minnesota survey					
Cities	1	2	3	6	
Countries	3	8	5	16	
State agencies	4	0	2	6	
Total	8	10	10	28	
Percentage	28	36	36		100

* *Notes*: Clearly open access criteria
- No charges or only costs of media/dissemination
- Price for entire database ≤ $1,000
- No restrictions on secondary use

Clearly revenue generation criteria
- Charges greater than costs of dissemination
- Price for entire database > $1,000
- Restrictions on secondary use

Mixed or conflicting policy sites
- Meet one or two revenue generation criteria

Source: Onsrud, Johnson and Winnecki (1996, Table 9)

based totally on allocating the costs of developing, maintaining and bundling data. At the other end of the spectrum, Portland (Oregon) Metro Service District is a good example of market demand pricing. Metro gave no consideration to the actual cost of developing the database when carrying out their initial market research and concentrated largely on actual demand and pricing sensitivities. Consequently the potential for cost recovery was only considered after they had developed a comprehensive understanding of market demand.

The extent of these activities throughout the United States is not known but the findings of two recent surveys of local and county GIS administrators in the country as a whole and in the state of Minnesota by Onsrud, Johnson and Winnecki (1996) show that 20 percent of the national sample and 36 percent of Minnesota respondents can be classified as clearly revenue generation sites (see Table 6.1). These satisfied three key criteria: they charged more than the dissemination and duplication costs for their data; they levelled fees in excess of $1,000 for their entire database; and they imposed restrictions on buyers with respect to the secondary use of their data. In addition a further 56 percent of respondents to the national sample and 36 percent of Minnesotans met one or two of these criterion but not all of them. In contrast only 24 percent of the national sample and 28 percent of the Minnesota respondents could be identified as clearly open access sites.

As a result of these developments it can be seen that there is a great deal of variation in marketing geographic information throughout the United States. While products such as the small-scale maps produced by the National Mapping Division of the United States Geological Survey and the census data collected by the Bureau of the Census are available at no cost or at most the marginal costs of reproduction, many states have modified their open records laws to restrict free access to geographic databases and some local government agencies are charging full market prices for large scale geographic information.

Legal Protection

Notwithstanding the general principles set out above governing public access to information collected by federal agencies, the intellectual property rights in digital databases are governed by the Copyright Act in the United States (Onsrud and Reis, 1995). The extent to which such databases can be protected by copyright law is now open to question following the decision reached by the US Supreme Court with respect to Feist Publications v. Rural Telephone Service in 1991. This case concerned a rural telephone company in Kansas which refused to license the use of its telephone directory to Feist Publications. Feist decided to extract the numbers it needed, claiming that they were facts and thus not subject to protection by copyright. The Supreme Court ruled with Feist, thereby overturning a long line of lower court rulings that favoured the 'sweat of the brow' position regarding the extent to which compilations of this kind are covered by copyright.

According to Karjala (1995, p. 396):

> the Feist decision relentlessly follows standard copyright dogma to a superficially unremarkable conclusion: copyright protects only expression, not fact; the expression protected must be the product of intellectual creativity and not merely labour, time or money invested; protected elements of the missing work are precisely those that reflect this intellectual creativity, and no more.

As a result, he concludes that confusion is the best description of the state of the art with respect to post-Feist copyright protection of digital geographic information. In his view the only legitimate remedy to this problem is either a statutory amendment of the 'compilation' definition or a *sui generis* database statute (Karjala, 1995, p. 415).

Matters relating to privacy at the federal level are governed principally by the 1974 Privacy Act. The Privacy Act

1. Allows individuals to determine what records pertaining to them are collected, maintained or used by federal agencies.
2. Allows individuals to prevent records obtained for a particular purpose from being used or made available for another purpose without their consent.
3. Allows individuals to gain access to such records, make copies of them and make corrections.
4. Requires agencies to ensure that any record which identifies individuals is for a necessary and lawful purpose, and
5. Requires agencies to provide adequate safeguards to prevent misuse of personal information. (Onsrud, Johnson and Lopez, 1994, p. 241)

Similar legislation also operates in many states to protect personal privacy. It should be noted, however, that while existing legislation imposes restrictions on the personal information that government may gather from private individuals it does not apply to the private sector (see Branscomb, 1994, p. 16 *et seq*, for a further discussion of these questions). Furthermore, it should also be noted that the extent to which privacy legislation has been enforced by public agencies has been criticised and that it has also been argued that adherence to privacy protection guidelines has not been a priority for federal agencies (Flaherty, 1989, p. 331).

ELEMENTS OF NATIONAL GEOGRAPHIC INFORMATION STRATEGY

Coordination

In 1989 the United States Office of Management and Budget (OMB) asked the Federal Interagency Coordinating Committee on Digital Cartography (FICCDC) to consider the need for expanding the scope of coordination with respect to spatial data usage. Its prime motives were to reduce the potential for waste and duplication of effort and exploit the potential offered by spatial data and related technologies for greater efficiency and effectiveness.

On the basis of the Committee's recommendations a revised Circular A-16 was issued by the OMB on 19 October 1990 entitled *Coordination of Surveying, Mapping and Related Spatial Data Activities* (OMB, 1990). This established an interagency Federal Geographic Data Committee to coordinate the 'development, use, sharing, and dissemination of surveying, mapping, and related spatial data' by

1. Promoting the development, maintenance, and management of distributed database systems that are national in scope for surveying, mapping, and other related spatial data;
2. Encouraging the development and implementation of standards, exchange formats, specifications, procedures, and guidelines;
3. Promoting technology development, transfer, and exchange;
4. Promoting interaction with other existing Federal coordinating mechanisms that have an interest in the generation, collection, use and transfer of spatial data . . .

(OMB, 1990, pp. 6–7)

One of the most important features of the revised Circular A-16 was its broad view of spatial data activities. The initial list of members of the FGDC included the Departments of Agriculture, Commerce, Defence, Energy, Housing and Urban Development, Interior, State, and Transportation as well as the Federal Emergency Management Agency, the Environmental Protection Agency, the National Aeronautics and Space Administration, and the National Archives and Records Administration. Since its establishment the Tennessee Valley Authority and the Library of Congress have joined the FGDC and any federal department with an interest in spatial data can request membership. The Committee secretariat of 12 staff is based in the National Mapping Division of the US Geological Survey in the Department of the Interior. The current chair of the FGDC is the secretary of the interior, Bruce Babbitt, who has a strong personal interest in these issues.

To carry out its tasks the Committee has created a framework consisting of 12 thematic subcommittees and four working groups shown in Figure 6.3. From this it can be seen that the responsibilities for leading coordination activities for different categories of data have been assigned to specific departments. For example, the Department of the Interior is the lead department for basic cartographic data and also for cadastral and geologic data while the Department of Agriculture has the responsibility for coordinating activities relating to soils and vegetation data and the Department of Commerce is responsible

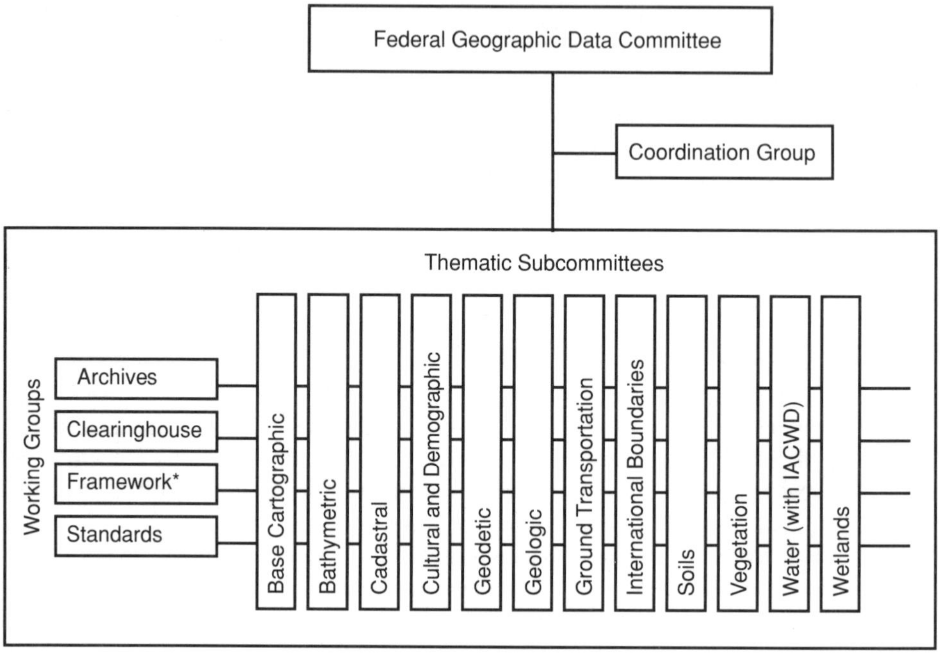

* - includes representatives of State and local government.

FGDC Subcommittees and Lead Departments

Base Cartographic - Interior
Bathymetric - Commerce
Cadastrat - Interior
Cultural and Demographic - Commerce
Geodetic - Commerce
Geologic - Interior
Ground Transportaiton - Transportation
International Boundaries - State
Soils - Agriculture
Vegetation - Agriculture
Water - cosponsored with the Interagency Advisory Committee on Water Data
Wetlands - Interior

Figure 6.3 Federal Geographic Data Committee structure.

for cultural and demographic data. The working groups deal with issues common to all spatial data categories with a view of promoting consistency among the subcommittees. The coordination group provides the means for the subcommittee and working groups to interact and coordinate their activities (Frederick, 1995).

At the same time as the release of Circular A-16 and the establishment of the FGDC, the Mapping Sciences Committee of the National Research Council started discussions about the national need for an infrastructure to promote the sharing of geographic information. The Mapping Sciences Committee (National Research Council, 1993) coined the phrase 'National Spatial Data Infrastructure' which was subsequently adopted by the FGDC to denote the national digital information resource embodied in Circular A-16. In 1993 the National Spatial Data Infrastructure was also listed as one of the initiatives to foster better intergovernmental relations in the National Performance Review initiated by Vice-President Gore (1993).

Four years after the establishment of the FGDC President Clinton signed Executive Order 12906 entitled 'Coordinating geographic data acquisition and access: the National Spatial Data Infrastructure' on 11 April 1994 'to strengthen and enhance the general policies described in OMB Circular A-16' (Executive Office of the President, 1994, Section 2a). This gave the FGDC the task of coordinating the federal government's development of the National Spatial Data Infrastructure and required that each member agency of that committee hold a policy-level position in their organisation. In this way the Executive Order significantly raised the political visibility of geospatial data collection, management and use not only among federal agencies but also nationally and internationally (Anderson, 1996; Tosta and Domaratz, 1996). The Executive Order also requires the FGDC 'to involve state, local and tribal governments in the development and implementation of the initiatives contained in this order' and to 'utilise the expertise of academia, the private sector, professional societies and others as necessary to aid in the development and implementation of the objectives of this order' (Executive Office of the President 1994, Section 2e).

The Executive Order defines the NSDI in the following terms: 'National Spatial Data Infrastructure (NSDI) means the technology, policies, standards and human resources necessary to acquire, process, store, distribute, and improve utilisation of geospatial data' (Section 1a).

The Executive Order also sets out in some detail the main tasks to be carried out and defines time limits for each of the initial stages of the development of the NSDI. Section 3 outlines the concept of the National Geospatial Data Clearinghouse which is defined as 'a distributed network of geospatial producers, managers and users linked electronically'. Section 4 sets out the procedures to be followed with respect to data standards while Section 5 of the Order describes the need to develop a National Digital Geospatial Data Framework. The Clearinghouse and Framework proposals are discussed in some detail later in this chapter with reference to subsequent developments.

An important feature of the Order is the emphasis that is given to the need for partnerships for data acquisition:

> the Secretary [of the Interior], under the auspices of the FGDC, and within nine months of the date of this Order, shall develop, to the extent permitted by law, strategies for maximising cooperative participatory efforts with State, local and tribal governments, the private sector, and other non federal organisations to share costs and improve efficiencies of acquiring geospatial data consistent with this order. (Section 6)

In this way the Executive Order establishes an environment within which new partnerships between the different levels of government and among other organisations are not only encouraged but required. In the eyes of the Secretary, Bruce Babbitt (1994, p. 31), this represents the greatest challenge for the FGDC:

> finding new ways to communicate more effectively and share resources among levels of government and between the public and private sectors is probably the greatest challenge facing us in the next year. It's my personal goal to see FGDC facilitate more enduring and productive partnerships for collecting, managing and using geospatial data to solve real problems.

With these considerations in mind, the FGDC has established a Competitive Cooperative Agreement Programme to help form partnerships that will facilitate the development and implementation of the National Spatial Data Infrastructure. This programme provides limited funding for cooperative projects with state and local government agencies, institutions of higher educations and private organisations. The first nine grants were made in 1994 mainly to promote the implementation of state nodes for the National Geospatial Data Clearinghouse (see below).

In essence, then, the National Spatial Data Infrastructure as outlined by the FGDC consists of a portfolio of policies, procedures, standards, technologies and organisations that lead to the more efficient use, management and production of geospatial data (Tosta, 1995, p. 106).

Core Data

The Executive Order required the FGDC to submit a plan within nine months for the initial implementation of a National Digital Geospatial Data Framework together with the necessary arrangements for ongoing data maintenance:

> The framework shall include geospatial data that are significant, in the determination of the FGDC, to a broad variety of users within any geographic area or nation-wide. And at minimum, the plan shall address how the initial transportation, hydrology and boundary elements of the framework might be completed by January 1998 in order to support the decennial census of 2000. (Executive Order of the President, 1994, Section 5)

To carry out this task a framework working group was set up consisting of representatives drawn not only from federal agencies but also including representatives nominated by state agencies as well as professional organisations such as the Urban and Regional Information Systems Association (URISA) and the American Congress of Surveying and Mapping (ACSM).

In its report to the OMB in March 1995, the FGDC (1995) set out the purpose and goals of the framework and defined the information content, as well as specifying the institutional roles required and outlining a strategy for its phased implementation. The purpose and goals of the framework are defined in the following terms:

> The framework is a basic, consistent set of digital geospatial data and supporting services that will:
>
> - provide a geospatial foundation to which an organisation can add detail and attach attribute information.
> - provide a base on which an organisation can accurately register and compile other themes of data, such as soils, or geology.
> - orient and link the results of an application to the landscape. (FGDC, 1995, p. 3)

The information content of the framework as defined in the report includes seven elements: geodetic control, digital orthoimagery, elevation data, transportation data, hydrography, administrative boundaries and cadastral information.

Since the publication of this report the main emphasis of FGDC activities has shifted to facilitating different kinds of geographic information management partnership at both the federal and the state and local government levels. With these considerations in mind, 24 case studies at different levels of government covering different kinds of geographic information are being undertaken to evaluate the current state of the art. The basic objectives of this evaluation are to identify examples of current best practice in this respect and to distil the lessons from this experience into a handbook for future partnership participants.

Metadata

The Executive Order required the FGDC to establish an electronic National Geospatial Data Clearinghouse for the National Spatial Data Infrastructure within six months (Executive Office of the President, 1994, Section 3a). In response to this demand, a report on federal agency activities was submitted in December 1994 which established roles and responsibilities, set out criteria for metadata documentation, and considered matters relating to the distribution and access of both metadata and geospatial data (FGDC,

1994b). In essence the National Geospatial Data Clearinghouse consists of three main components: metadata, the Internet and distributed search and query tools. Together these provide the ability to document, serve, search for, browse and access geospatial data that vary in scope, context, format, detail and location (Tosta, 1995, p. 110).

The FGDC has developed its own metadata content standards to describe the quality and characteristics of geospatial data. The content standards define data elements for the following topics:

- Identification information: e.g. title, geographic area covered and currentness.
- Data quality information: e.g. positional attribute accuracy, completeness, consistency, methods used to produce data.
- Spatial data organisation information: e.g. methods used to represent spatial positions directly (such as raster or vector) and indirectly (such as street addresses or county codes).
- Spatial reference information: e.g. the name and parameters of map projections or grid coordinate systems and the coordinate system resolution.
- Attribute information: e.g. the names and definitions of features, attributes and attribute values.
- Distribution information: e.g. distributor contact, available formats, information on how to obtain data sets on-line or on physical media and fees for the data.
- Metadata reference information: e.g. currentness and information about the organisation that provided the data.

Data producers throughout the United States are being encouraged to describe the data that they produce using this standard and all federal agencies have been required to use this standard since 1995 to document their new data.

The second component of the Clearinghouse is the Internet. The Internet is a global network of networks that enables computers of all kinds to communicate and share information throughout the world. Given the opportunities for dissemination that are being opened up by this technology, federal agencies are being encouraged to establish Internet connections and create indexed databases for access to their holdings.

The last component of the Clearinghouse is the development of software tools for searching and querying data on the network. According to Tosta (1995, p. 110), the FGDC has been testing Wide Area Information Services (WAIS) software which is built on a library standard (Z39.50) for search and queries. WAIS was originally designed for text search but has been enhanced to enable geographic searches based on the specification of the coordinates covering the area of interest, and further software development is currently under way to enable geographic searches based on irregular polygons defined by customers.

By mid-1996 there were 11 fully active nodes that provided an indexed database and hundreds of home pages describing data in the Clearinghouse. A similar number of other nodes were under development. These include USGS and other federal agency nodes relating to specific types of government activity as well as state-level nodes covering a wide range of users. A good example of the latter is the North Carolina state node which was referred to earlier in the chapter. This provides information about the data holdings of its six current participants: the Centre for Geographic Information and Analysis; the City of Charlotte/Mecklenberg County, which is the largest urban area in the state; the North Carolina Department of Environment, Health and Natural Resources division of Coastal Management, which covers 20 coastal counties in the state; the North Carolina

Department of Environment, Health and Natural Resources division of Water Resources, which works closely with federal agencies; the Triangle J Council of Governments in the Raleigh, Durham and Chapel Hill area; and the University of North Carolina's Institute for Transportation Research and Education.

EVALUATION

Given the number of interests involved, the development of national geographic information strategies in the United States is a mammoth task which is made particularly difficult as a result of the devolution of responsibilities to state and local governments under the American government system. As a result, as many as 80,000 agencies are involved in some way with geographic information in the public sector alone. In addition to this it is important to recognise that many private sector companies in the utilities fields and that of land title insurance hold large quantities of data that would be held by public or semi-public agencies in countries such as Britain, the Netherlands and Australia.

Despite these difficulties, the situation in the United States also has its advantages. As a result of the large number of interests involved, there is a long tradition of information sharing and coordination at state and local government levels. Consequently a great deal of operational experience has been built up not just at these levels but also as a result of the activities of professional organisations such as the American Congress of Surveying and Mapping and the Urban and Regional Information Systems Association over the last quarter of a century.

Nevertheless there are wide variations between states and even between different areas of the same state in terms of geographic information handling practice. This is also the case with federal agencies. In addition to the US Geological Survey many federal agencies have major holdings of geographic information and there are also considerable federal land holdings in many states.

Despite these difficulties, the collaborative ventures between the USGS and the Census Bureau to develop, first, the GBF/DIME files for the 1970 and 1980 censuses of population, and second, a comprehensive set of TIGER files for the 1990 and 2000 censuses, represent a win–win situation for both agencies as a result of greater coordination of geographic information. The residual cost/copyright free dissemination policies followed by both these agencies with respect to these data have also provided a major boost to the development of value-added products in the private sector of the national economy.

The need for greater coordination of geographic information holdings at the federal level was recognised by the creation of the Federal Geographic Data Committee in 1990. The mandate of the Federal Geographic Data Committee was subsequently extended as a result of the Executive Order setting up the National Spatial Data Infrastructure in 1994. This increased the political visibility of existing federal and state-wide coordination efforts.

One of the most impressive features of the work of the Federal Geographic Data Committee is the degree of political support for its activities. This originates at the highest levels of government as a result of Vice-President Gore's involvement in the information superhighway project and the fact that the FGDC is chaired by the secretary of the interior. The possibility that budget penalties might be imposed by the Office of Management and Budget on federal agencies that do not meet their statutory requirements with respect to geographic information is also a potential impetus to building the NSDI.

Nevertheless it must be recognised that the implementation of the NSDI requires a collaborative effort on the part of large numbers of separate agencies and this can only be achieved by the creation of partnerships on a hitherto unprecedented scale not only between different federal agencies, but also between federal agencies and state/local government agencies and between these agencies and the private sector. In practical terms this represents the main challenge for those involved in the creation of the NSDI.

With this in mind, it is not surprising therefore that the need to create partnerships and involve all the key players is evident in current US efforts to implement a National Spatial Data Infrastructure. This can be seen in the complex matrix management structure developed by the FGDC which requires the participation of representatives from an unusually wide range of public and private agencies. It is also apparent in the activities of the National Geospatial Data Framework group which is addressing the core data development programme and the decision to implement a decentralised version of a National Geospatial Data Clearinghouse which once more involves many different kinds of organisation.

Despite these developments there are still major problems to be resolved with respect to access to data held by public agencies in the United States. This is particularly the case at the state/local government level following the changes that have been made to open records laws in some states to restrict public access and the efforts of a large number of local government agencies to recover some or all of the costs of their operations through the sale of their data. These tendencies run counter to those contained in the latest version of OMB Circular A-130 for federal agencies which reiterates the principles enshrined in the Freedom of Information Act. Such developments not only reflect the tensions that exist between different types of public agencies with respect to access to information, but also present differing views regarding the ownership of information.

How far the United States will be able to implement the National Spatial Data Infrastructure is bound to be a matter for conjecture, given the large number of interests involved and the extent to which it is inherently dependent on essentially fragmented and voluntary efforts to promote data sharing on the part of federal, state and local governments. Nevertheless, the impetus that has been built up as a result of recent efforts is very impressive, as is the political support behind them. For this reason, it is worth considering the vision of the secretary of the interior, Bruce Babbitt (1996):

> I can only speak from experience, of course. But as we used GIS as a tool to approach the complex challenges of Glen Canyon [in the Grand Canyon], that experience revealed three seminal lessons in how we, as a nation, shall re-establish strong, lasting, nourishing roots in our ever changing landscapes of complexity.
>
> The first lesson is how GIS empowers us to see our landscape in an entirely new spatial dimension. We see not fragments – structures, roads, minerals, animals, plants, water and soil – but the whole watershed as one interconnected unit . . .
>
> But you cannot see the intricate complexity within each watershed from the banks of the Potomac. You need a lens that fits the needs of each specific site. Which leads us to the second lesson from Glen Canyon: sound decisions can only come through good science that is informed by local stakeholders. This lesson forces us to reverse the traditional flow of information – from the top down – and reconsider how that information is collected and used at the local level . . .
>
> The final, and perhaps the most important lesson from the Glen Canyon experience is that GIS plays a critical role by helping inform a complex democratic society such as ours . . . GIS integrates that information and enables every stakeholder to use it.

CHAPTER SEVEN

What Lessons Can Be Learnt From These Experiences?

INTRODUCTION

The final chapter of this book compares and evaluates the findings of the four country case studies with a view to identifying what general lessons can be learnt from these experiences in connection with the development of national geographic information strategies. In the process reference is repeatedly made to the various conceptual standpoints regarding the nature of geographic information that were discussed in Chapter 2. The basic structure of the presentation follows that of the case studies themselves, with separate sections describing the activities of the main providers of geographic information, the institutional context within which national geographic information strategies are formulated, and the three key elements of national geographic information strategy defined in Chapter 2. The final section of the chapter draws a number of general lessons from this experience that need to be taken into account by other countries in the development of their national geographic information strategies.

Although the four case-study countries have many features in common, there are also considerable differences between them with respect to their history and geography which have important implications for the development of their national geographic information strategies. Both Britain and the Netherlands have continuous settlement histories dating back several thousand years. In contrast, the modern settlement history of both Australia and the United States dates back only several centuries and large parts of both countries have been colonised in the last hundred years.

Some statistics relating to land area, population size and population density for the four countries are contained in Table 7.1. From this it can be seen that there are major differences between them with respect to all three criteria. It shows, for example, that the land area of the largest country, the United States, is over 200 times that of the smallest country, the Netherlands. However, this difference is reduced to 20 times when the population sizes are compared as the density of population in the Netherlands is nearly 15 times that in the United States.

Generally the United Kingdom has more in common with the Netherlands in these respects than with either Australia or the United States. It covers a land area six times that of the Netherlands but, because of its lower population density, has a population size

Table 7.1 Case-study countries: land area and population

	United Kingdom	Netherlands	Australia	United States
Area (000 km^2)	244.1	40.8	7686.8	9809.1
Population (000)	56470	15240	17746	259681
Density (pop./km^2)	231.3	373.5	2.3	26.5

that is only four times greater that of the Netherlands. With a land area that is four-fifths that of the United States, Australia has more in common with that country than with the other two countries. However, it also differs from it markedly with respect to population size because of its very low density of only 2.3 persons per square kilometre.

MAIN PROVIDERS OF GEOGRAPHIC INFORMATION

In all four case-study countries a wide range of government agencies are involved in the collection and dissemination of geographic information. It has been necessary to leave out many of these from the detailed analysis for practical considerations and to concentrate the discussion around the activities of three main types of geographic information provider: agencies with responsibilities for land titles registration, surveying and mapping agencies, and organisations concerned with socio-economic data provision.

Land Titles and Cadastral Information

The findings of the four case studies show pronounced differences between them with respect to both the degree of centralisation of land titles administration and the comprehensiveness of coverage which have important implications for their role as geographic information providers.

Of the four land titles administrative systems, the Dutch Cadastre is the most centralised and comprehensive in coverage. It holds details of all 7 million land parcels in the country as well as 3.5 million owners. Unlike most other cadastral organisations it combines large-scale surveying and mapping tasks with land records administration activities in a single organisation.

Land titles administration is also highly centralised and comprehensive in coverage in Australia at the state and territory level although there are eight separate systems in operation in Australia as a whole. The number of parcels held by the states and territories varies considerably from over 3 million in New South Wales to only 50,000 in the Northern Territory. Unlike the Netherlands, however, surveying and mapping tasks are carried out in most states by the surveyor general's department whereas responsibility for land titles registration is given to the registrar general or the director of the land titles office.

Britain, rather England and Wales, also has a centralised land titles registration system which is administered by HM Land Registry. This holds about 16 million records, but is not fully comprehensive as its records are limited mainly to parcels where transactions have taken place since the Land Registration Act was passed in 1925. The functions of the Land Registry are limited to land titles registration and it is required by law to record its data holdings on Ordnance Survey maps.

Unlike the other three countries, land titles records responsibilities are devolved in the United States to local government and the ways in which these are discharged varies considerably from state to state and even within areas of the same state. In most cases there is also a split between land titles administration and surveying and mapping responsibilities. Given these circumstances it is not surprising that the desire to coordinate such activities, together with the need to provide both technical and financial assistance to local government has featured prominently on the agendas of state legislatures for the last twenty years.

National Mapping Agencies

From the four case studies it can be seen that there are also marked variations between the national mapping agencies in the four countries with respect to their range of responsibilities, their size, the extent to which they have created digital databases and the products that have been developed from these databases.

Ordnance Survey of Great Britain stands out from the other three national mapping agencies because its duties include both large- and small-scale mapping for the country as a whole. Elsewhere the national mapping agencies are responsible only for small-scale mapping down to the 1:10,000 scale in the case of the Dutch Topografische Dienst, 1:24,000 (or 1:12,000 if the Digital Orthophoto Quadrangles are included) in the case of the National Mapping Division of the US Geological Survey, and 1:100,000 in the case of the Australian Surveying and Land Information Group (AUSLIG).

These differences are reflected in the size of each agency once the area of the country is taken into account. Ordnance Survey is the largest national mapping agency with about 2,000 employees, followed by the US National Mapping Division with 1,500 employees. Both the Dutch and Australian agencies are very small by comparison.

The core activity of most national mapping agencies is the maintenance of the national topographic database in a form that meets present and future demands. In May 1995 Ordnance Survey completed the digitisation of nearly 230,000 map sheets covering the land area of Great Britain and now provides an accurate and up-to-date digital topographic database for the whole country. AUSLIG in Australia has national digital coverage at the 1:250,000 scale, while the Dutch Topografische Dienst plans full digital coverage at the 1:10,000 scale by the end of 1997. In the United States digital line graphs are available for the country as a whole from the National Mapping Division for two key features, transportation and hydrography, as a result of the agreement reached with the Bureau of the Census in connection with the 1990 Census of Population.

Ordnance Survey has developed a wide range of digital products from the national topographic database together with other databases. These include products such as ADDRESS-POINT and OSCAR which are derived essentially from the large-scale maps. It should also be borne in mind that these products often reflect the position that is occupied by the national mapping agency within government. In Britain, Ordnance Survey is an Executive Agency with a considerable degree of autonomy. In contrast the Dutch Topografische Dienst is part of the Ministry of Defence and the US National Mapping Division is an integral part of the US Geological Survey within the Department of the Interior, while AUSLIG is part of the Commonwealth Department of Administrative Services.

Socio-economic Statistics

Unlike land titles registration and national mapping responsibilities, the arrangements for statistical data collection in the four case-study countries are similar in all of them. However, there are some important differences between them with respect to the arrangements for census-taking and the extent to which these agencies exploit the potential of census geographies.

The British Office for National Statistics, Statistics Netherlands, the Australian Bureau of Statistics and the US Bureau of the Census are all large and complex organisations with responsibilities for the collection of both demographic and economic information. It should be noted, however, that in Britain many other government ministries such as the Department of the Environment also collect statistics on their own behalf under the overall auspices of the Government Statistical Service. In Australia it should also be borne in mind that the majority of the 3,500 people employed by the Australian Bureau of Statistics work in branches in the states and territories where they fulfil state-wide as well as Commonwealth responsibilities.

One of the most important statistical products from a geographic information point of view in most countries is the census of population. Over the last 30 years the US Bureau of the Census has developed into a primary user of both geographic and attribute data. Together with the National Mapping Division of the USGS it has played a major role in pioneering large-scale mapping in cooperation with local governments and private companies through the creation of GBF/DIME and TIGER files since the 1970 Census. The Australian Bureau of Statistics broke new ground with respect to the integrated digital map base that was created for census design, mapping and collection purposes in the 1996 Census of Population and Housing by a consortium of public sector mapping agencies. This has provided a major boost to the creation of a national digital database for Australia as a whole. Similarly the former British Office of Population Censuses and Surveys has worked closely with other key players in the geographic information field in the planning of its decennial censuses and the development of geographic information products based on census data.

Given the importance attached to census-taking in these three countries, it may be surprising to find that there has been no census in the Netherlands since 1971. However, this is because the population registers (GBA) maintained by each municipality contain details about every individual in the country and it is argued that their quality and accuracy is such that they make conventional census-taking unnecessary.

Some Conclusions

The findings of the analysis show that there are marked differences in the distribution of responsibilities between the different levels of government in the four case-study countries. Table 7.2 summarises these differences. From this it can be seen that Britain and the Netherlands are relatively centralised with respect to responsibilities for land titles registration, mapping and socio-economic data provision. In contrast, in the federal systems of Australia and the United States, many of the responsibilities carried out centrally in Britain and the Netherlands are carried out at the state or territory level. In Australia, for example, land titles registration and most surveying and mapping activities are carried out centrally at the state and territory level, while responsibility for socio-economic data lies at the Commonwealth level. From the standpoint of geographic information, the

Table 7.2 Case-study countries: distribution of responsibilities among different levels of government

	Britain*	Netherlands	Australia	United States
Central government	Land titles registration, small- and large-scale mapping, statistical data	Land titles registration, small- and large-scale mapping, statistical data	Some small-scale mapping, statistical data	Small-scale mapping, statistical data
State/territory government	N/A	N/A	Land titles registration, small- and large-scale mapping	Some land titles registration and small- and large-scale mapping
Local government	None	Some large-scale mapping, population registers	Some large-scale mapping	Land titles registration, large-scale mapping

* The term 'Britain' in this case refers to England and Wales although the responsibilities of some agencies (e.g. Ordnance Survey) cover Great Britain as a whole.

United States has an even more decentralised structure with land titles registration and large-scale surveying and mapping responsibilities devolved in most states to the county or municipal level.

INSTITUTIONAL CONTEXT

The findings of the four case studies highlight the differences that exist between them with respect to the general institutional context which governs geographic information provision. Of particular importance in this respect are various factors governing the dissemination of geographic information including the role of government, pricing and cost recovery and the need for regulation, and a number of matters concerning the legal protection of digital databases including copyright and the need to protect personal privacy.

Disseminating Geographic Information

The Changing Role of Government

In all four case-study countries there have been wide-ranging major changes in the role of government over the last few years which have important implications for the dissemination of geographic information. *Reinventing Government*, as the title of Osborne and Gaebler's (1992) classic book suggests, involves reassessing the operations carried out by governments with particular reference to privatisation, deregulation and market testing. An important consequence of this is that government agencies are increasingly expected to operate in a more commercial way. These general trends have had a particularly marked impact on the main providers of geographic information in the four case-study countries because they come at a time when they face growing demands from users for high quality data as a result of the introduction and use of geographic information technologies.

The effect of such developments is particularly apparent in Britain as a result of the efforts of successive Conservative governments to reinvent government since 1979. Since the launch of the Next Steps Programme in 1988, the main government providers of geographic information have been required to deliver services more efficiently and effectively within available resources. In the process HM Land Registry, Ordnance Survey and the Central Statistical Office have become Executive Agencies and HM Land Registry also has Trading Fund status. Each of these Executive Agencies is given financial, efficiency and customer service targets so that their performance over time can be closely monitored.

At the same time, British government agencies have been encouraged to contract out work to private sector agencies and also to participate in joint ventures with them. Contracting out played an important part in the creation of the national digital topographic database and Ordnance Survey has also been active in a number of joint ventures with private sector companies. In the process its work has become increasingly customer driven, as indicated by its mission statement 'to be the customer's first choice for mapping today and tomorrow'.

A similar pattern of developments can be found in the Netherlands. The Cadastre became an independent administrative body (ZBO) in 1994. The secretary of state of the Ministry of Housing, Spatial Planning and the Environment must approve its long-term policy plans and the fees that it charges but in every other respect it operates as an independent organisation. Under the terms of its Foundation Act the Cadastre is also allowed to carry out additional activities, provided that they are not subsidised by its core business and are approved by the secretary of state. As a result, it set up a separate company in April 1996 to develop a new postal address coordinate data product for the whole of the Netherlands. Like Ordnance Survey, the Cadastre is increasingly customer driven and its current activities have been described by its director of land information, Paul van der Molen (1996), as 'monopoly in a customer driven way'.

Statistics Netherlands was substantially restructured in 1993 into a flatter organisation consisting of a number of divisions which have been given a great deal of autonomy. Underlying this restructuring is the desire to transform Statistics Netherlands from 'a factory of figures to a node on the electronic highway'. One consequence of these changes is an increasing number of joint ventures with commercial value-added resellers.

In Australia the impact of such developments is particularly marked at the state and territory level through the establishment of agencies such as the Office of Geographic Data Coordination in Victoria to provide a major capital asset for the state through its spatial data infrastructure. A key component of this agency's mandate is to develop a financial model and pricing policy to achieve sustainable maintenance, return on capital and industry stimulation.

At the Commonwealth level, the Australian Surveying and Land Information Group (AUSLIG) operates as a separate business unit within the Department of Administrative Services, providing a wide range of services for both public and private sector clients in the fields of mapping, geodesy, surveying and remote sensing. Its existing commercial activities are currently being tested for sale through a management buyout or trade sale. The Australia Bureau of Statistics has also developed a wide range of statistical products in conjunction with private sector agencies and also provides a variety of customised services based on its demographic and business databases.

The National Mapping Division of the US Geological Survey is also having to reconsider its activities as a result of financial pressures. The size of its appropriation from the federal government has declined in real terms over the last few years and its capacity to generate additional revenue through the sale of its products and services is limited, as most of its customers are federal agencies which face similar pressures. The Division is also being required to contract out an increasing proportion of its mapping activities to private sector companies. It is anticipated that this will rise from 24 percent in 1995 to 60 percent by the end of 1998.

Pricing and Cost Recovery

A key factor with respect to the dissemination of geographic information in the context of the changes that are taking place in the role of government is the extent to which the pricing of geographic information products is governed by considerations of cost recovery. As pointed out in Chapter 2, geographic information has many features of a public good in economics, i.e. its benefits can be shared by many people without loss to any individual and it is not easy to exclude people from these benefits. In practice, however, as the discussion of geographic information as a commodity showed, it is possible to exclude people to some extent from these benefits by means of price, but when such measures are employed by government agencies, this runs counter to the needs for open government and public accountability.

These tensions are closely reflected in the practices of the four case-study countries with respect to pricing and cost recovery. The United States differs from the other three countries in that federal government agencies are required by law to make the information they collect available to the public at no more than the marginal costs of dissemination and with no copyright restrictions. This is the policy followed by the National Mapping Division of the US Geological Survey with respect to the distribution of its map products and also by the US Bureau of the Census with respect to its census products. However, the position at the state and local government levels with respect to the dissemination of geographic information products is much less clear. Some states have recently modified their open records laws to restrict free access to their geographic databases and a large number of local government agencies have taken steps to generate revenue through the sale of their products and/or to impose restrictions on buyers with respect to the secondary use of their data.

Elsewhere the position with respect to pricing and cost recovery depends very much on the nature of the agency. In Britain agencies such as the Office for National Statistics, which collect data primarily for government policy-making, are characterised by low levels of cost recovery, whereas agencies such as HM Land Registry, which collect information primarily for regulatory purposes, are obliged to recover their costs fully. Agencies where there is a public good or national interest component such as Ordnance Survey fall somewhere between these extremes.

Similarly in the Netherlands, Statistics Netherlands falls into the first category while the Cadastre comes clearly into the regulatory agency category, with the Topografische Dienst coming in the intermediate category. In Australia the position is more complex because of the nature of state land administration requirements. Nevertheless, even in this case, the Australia Bureau of Statistics falls into the first category.

In all three countries there have been attempts to distinguish between the public good and non-public good components of government information products. A good example is the draft national policy for the transfer of land-related data in Australia which was approved by ANZLIC in October 1993. This distinguishes between data collected and funded in the public interest which should be made available at the average cost of transfer and further activities which should be subject to additional charges above and beyond recovering the full costs of distribution. Similarly in Britain, government guidelines for charging for statistical services and products make it clear that while the costs of collecting, processing, analysing and presenting statistical information for official use should be met from public funds, information published for wider use should normally be priced to recover the full cost of making that information available in excess of that required for government use.

In practice, however, the implementation of such principles is not straightforward. In Australia, for example, despite the draft agreement on the transfer of land-related data, pricing policies vary considerably from state to state and from territory to territory with respect to the use of digital cadastral databases by non-government agencies with some states setting their prices low to encourage users and others trying to recover some of the additional costs of data capture and maintenance in their pricing. There is also the question as to how far development costs should be taken into account, as McLennan (1995, p. 20) has pointed out with respect to the needs of the Australian market for statistical information:

> The ABS needs to invest resources in the development, production and delivery of such products and services, and the price of such products and services are set at market prices where quantifiable, but at least to recover the full costs involved, including amortisation of capital and a reasonable allowance for contingencies and risk involved.

Regulation

One consequence of the changes that have been taking place in governments generally and the privatisation of natural monopolies such as electricity, water and gas supply is that this has created a need to regulate the prices charged by such agencies and the access to their facilities in the public interest. As the main geographic information providers in government have some of the features of natural monopolies, questions have also been raised as to whether there is a need to regulate their activities. It has been argued, for example, that the pressures on public sector agencies in Australia to cover some or all of their costs of their non-public interest activities makes it necessary to establish access rights and rules in the context of national competition policy (Price Waterhouse, 1995, p. 53). On the other hand, Coopers and Lybrand (1996, p. 8) argue that the case for price control regulation in Britain is relatively weak on the grounds of the small size of the information providers relative to bodies such as the utilities and the sophistication of the key users of these products and services.

Legal Protection

Copyright

Despite Branscomb's (1995, p. 17) argument that traditional measures for controlling the flow of information need rethinking in the context of digital information, copyright is still the main instrument used for protecting intellectual property rights in electronic databases in the four case-study countries.

Britain takes the hardest line of the four countries on copyright. Under the Copyright, Designs and Patents Act of 1988 it is enough for compilers to show that they have expended work and effort to create digital databases (sweat of brow) without necessarily having to demonstrate that this involved original and creative effort. As a result, most geographic information products in Britain are covered by the provisions of copyright law, provided that they required skill and labour and have not been copied.

Australian copyright law follows similar provisions to that of Britain. Under the 1968 Copyright Act (and subsequent amendments) digital databases are protected, provided that 'judgement, skill and labour' have been expended on their creation.

In the United States small-scale maps and census data are provided by the federal agencies concerned without any restrictions on copyright to meet the requirements of the Freedom of Information Act. The extent to which other databases are protected by copyright law is open to question since the Supreme Court ruling in 1991 with respect to Feist Publications v. Rural Telephone Service. In this case the court took the view that compilations of facts such as telephone directories are not covered by the law. In effect, then, US copyright law protects expression, not facts, provided that expression is the product of intellectual creativity and not merely labour, time or money invested (i.e. sweat of the brow).

The position in the Netherlands is similar to the post-Feist position in the United States. Under Dutch copyright law, facts are not protected in their own right because they are not regarded as original, and the amount of time and effort that goes into the creation of the work is not seen to be relevant. To qualify for copyright protection, works must have a character of their own and/or bear the personal stamp of the author. As a result, it has been argued that copyright in its current form is not a suitable instrument for giving geographic databases legal protection in the Netherlands.

Both Britain and the Netherlands are members of the European Union and they are also required to implement any Directives agreed by the European Commission. In March 1996 a Directive on legal protection of databases was finally agreed after protracted negotiations. This has important implications for copyright provision and may in time bring Britain more in line with other European countries such as the Netherlands on these matters (see van Eechoud, 1996, for a comparison of copyright in the EU countries).

Data Protection

All four countries have data protection and/or privacy legislation designed to regulate the use of digital information relating to individuals. The British Data Protection Act, for example, requires all users to register and abide by its principles to ensure that personal data are obtained and processed fairly and used solely for previously specified purposes. Similarly the Dutch Personal Records Act specifies the contents of and access to personal records required for the protection of privacy, while the US Privacy Act enables individuals to prevent records that have been collected for one purpose being used for another purpose without their consent.

Despite these good intentions, however, the scope and extent of enforcement of data protection legislation in the case-study countries has been criticised. American legislation imposes restrictions on the use of personal information collected by public agencies but does not apply to the private sector. It has also been argued that the enforcement of such legislation leaves something to be desired.

ELEMENTS OF NATIONAL GEOGRAPHIC INFORMATION STRATEGY

The discussion of geographic information as an asset in Chapter 2 highlighted the need for governments to take steps to manage their geographic information assets in the national interest and exploit the opportunities opened up by the development of geographic information systems technology. In the process it was found that two different kinds of asset management are involved: the custodianship of particular databases and the coordination and management of the national portfolio of geographic information assets. As a result, two of the three key elements of national geographic information strategy were identified: the need for some measure of national coordination and the need for the development of metadata services to enable users to locate the individual databases they require within a distributed network.

The discussion of geographic information as an infrastructure drew attention to the extent to which national geographic information infrastructures are not just a collection of databases but also the people, the technology and the cocktail of laws, precedents and procedures described by Rhind (1996a). It was also noted that these infrastructures already exist and that the primary task facing governments is to develop national geographic information strategies to exploit the potential offered by GIS technology. An important element of any infrastructure of this kind is the core geographic databases that are used for a large number of different applications. Because of their overall strategic importance it was argued that these core databases constitute the final key element of any national geographic information strategy.

In practice, as the findings of the four case studies show, coordination and the establishment of core databases have taken up most of the efforts associated with the development of national geographic information strategies. In contrast the development of metadata services is a technically challenging but less politically demanding task. Consequently the bulk of the following discussion is devoted to the first two activities.

Coordination

In three of the four case-study countries, special bodies have been established with responsibilities for national geographic information coordination. The Dutch Council for Real Estate Information (Ravi) was set up in 1984 to advise the minister for housing, spatial planning and the environment on matters relating to the operations of the Cadastre. It was restructured as a national council for geographic information in 1993. The Australia Land Information Council (ALIC) was established in 1986 by agreement between the Australian prime minister and the heads of the state governments to coordinate the collection and transfer of land-related information between the different levels of government. In 1991 New Zealand become a full member of the council which was renamed the Australia New Zealand Land Information Council (ANZLIC). The Federal Geographic Data Committee (FGDC) was set up in 1990 as an interagency federal committee to coordinate the development, use, sharing and dissemination of surveying, mapping and related spatial data. As a result of the Executive Order of the President in 1994 its responsibilities were extended with respect to its work on the National Spatial Data Infrastructure. There are obvious parallels between these initiatives and the recommendations made to the secretary of state for the environment by the Chorley Committee in Britain regarding the establishment of an independent Centre for Geographic Information with strong links to government (Department of the Environment, 1987). In the view of the Chorley Committee one of the main tasks of this centre was to prepare proposals for developing national policy with respect to the availability of government-held spatial data and the development of spatial referencing systems and standards to facilitate the integration of geographic information.

The British government took the view that a centre of this kind was not needed and that it was better to encourage existing organisations to expand their range of activities than to set up new bodies. As a result, despite the subsequent establishment of an Association for Geographic Information with government support, Britain still lacks a formal national coordinating body which is comparable with those of the other three case-study countries. Nevertheless, some progress has been made with respect to coordination. The Chorley Report has stimulated a considerable degree of collaboration between government departments. This is reflected in the activities of the Interdepartmental Group on Geographic Information (IGGI) which regularly brings together representatives from over 30 government agencies to discuss issues relating to coordination. It should also be noted that, in the absence of a formal body, Ordnance Survey has played an important role in the development of national geographic information strategies and that its director general is the official adviser to the UK government on all survey, mapping and GIS matters.

Of the three formal national coordination bodies in the case-study countries the Federal Geographic Data Committee (FGDC) has the most impressive political profile. It is chaired by the secretary of the interior and senior-level representation is required as a result of the Executive Order signed by President Clinton. It also takes a particularly broad view of spatial data activities and its membership includes representatives of the Departments of Defense and State as well as the Departments of Commerce and Interior which house the Bureau of the Census and the US Geological Survey. This is particularly apparent in the topics covered by its thematic committees which range from basic cartographic data to transportation issues on the one hand and from soils and vegetation data to cultural and demographic information on the other.

The Australian coordinating body, ANZLIC, differs in several important ways from the FGDC. It is primarily concerned with coordinating the transfer of geographic information *between* the different levels of government. Consequently, only one of its ten members comes from a Commonwealth agency, eight represent the interests of the individual states and territories, and the remaining member represents the interests of New Zealand. From this structure it can be seen that ANZLIC can be regarded as a higher-level coordinating body of other coordinating bodies. Its Commonwealth member, for example, represents the interests of the Commonwealth Spatial Data Committee which is itself the coordinating body for the different Commonwealth ministries, while the state and territory representatives represent the interests of state and territory coordinating bodies such as the South Australia Land Information Council.

Given the distribution of responsibilities between the different levels of government in Australia and the extent to which primary surveying and mapping activities are devolved to the states and territories, it can be argued that a structure of this kind is essential to ensure the direct involvement of the main providers of geographic information. In this way the formal structure of ANZLIC builds directly on the coordination efforts of the states and territories in contrast to the FGDC in the United States.

On the other hand, it can be argued that ANZLIC takes a more restricted view of geographic information than the FGDC. Its primary responsibilities are with respect to the coordination of land information management activities, particularly with respect to surveying and cadastral responsibilities as, unlike the United States, these two responsibilities are centralised at the state and the territory level. Nevertheless, it should also be noted that its responsibilities, like those of most state and territory land information councils, also cover social economic data, environmental and natural resource information and materials related to the operations of the utilities.

Unlike either the FGDC or ANZLIC, the Ravi is an independent body which periodically reports to the secretary of state for housing, spatial planning and the environment. Its membership includes most (but not all) the main data providers and users in the public sector in the Netherlands in both central and local government. Unlike either of the other two bodies it also functions as a business platform to facilitate consultations with the business world and to inform the Dutch geographic information industry about its projects. The Ravi has no formal powers to compel public or private sector companies to participate in its activities and is consequently very much dependent on the enthusiasm and willingness of its members to support its efforts. Nevertheless, given the extent of intragovernment cooperation over the last few years, this essentially pragmatic approach has paid rich dividends.

The Ravi also differs from the FGDC and ANZLIC in another important way. It has no base within a government department and its future funding is therefore dependent on the continuing willingness of the secretary of state to provide some core funding for its activities and the extent to which its members are willing (and able) to supplement this core funding.

One feature that all three coordinating bodies have in common, however, is that they are modest in their staffing requirements. Even the largest body, the FGDC, has a staff of only 12, while ANZLIC and the Ravi make do with only four. These numbers indicate the scale of support operations needed to service the work of their members.

There are also close parallels between the vision of the National Spatial Data Infrastructure that is set out in the Executive Order signed by President Clinton, the latest strategic plan published by ANZLIC and the Ravi discussion paper on the National Geographic Information Infrastructure. All three documents identify similar priorities for coordination such as the need for agreed standards for data exchange and common spatial referencing systems. Furthermore, all three documents identify the strategic importance of creating core databases and the need for metadata services (see below).

From this analysis it can be seen that there are important differences between the three formal national coordinating bodies. The FGDC has a much higher political profile than either of the other bodies. On the other hand, the structure of ANZLIC takes direct account of the distribution of responsibilities between the different levels of government in Australia. Both the FGDC and ANZLIC operate from within government whereas the Ravi maintains strong links with government while keeping its independence. For this reason it comes closest to the model put forward for Britain by the Chorley Committee and rejected by the British government in 1988.

Core Data

There is no clear consensus among the four case-study countries with respect to the definitions of core data despite the general recognition of its strategic importance. Probably the best definition of the contents of a core data set is that contained in the National Digital Geospatial Framework document prepared by the FGDC Framework working group in the United States. The information content of the framework as defined in this document includes seven elements: geodetic control, digital orthoimagery, elevation data, transportation data, hydrography, administrative boundaries and cadastral information.

The national digital topographic database completed by Ordnance Survey of Great Britain in May 1995 contains all these elements with the exception of cadastral information. Ordnance Survey has also set out its vision of how a National Geospatial Data

Framework might be created by linking its national topographic database to spatial data held by other government departments. This will not be the property of one organisation but 'the totality of many individual data sets collected and held separately by many different organisations'. Some indication of how this objective might be achieved is given by the experience of the pilot project to explore the feasibility of establishing a National Land Information Service. This linked together land titles registration data held by HM Land Registry, property valuation data held by the Valuation Office, and planning data held by the city of Bristol to the national topographic database held by Ordnance Survey. The findings of this pilot project indicate that it is technically feasible but the business case for a National Land Information Service has still to be accepted by government.

Because of the division of responsibilities in Australia there are eight separate systems of land titles registration and eight separate surveying and mapping systems in operation although there are many similarities between them. Taken together, these digital cadastral databases and these digital topographic databases meet all the requirements set out above for core data. Apart from Tasmania, all of them are already available in digital form and are being incorporated in larger multipurpose databases by many state land information coordination bodies as can be seen from the examples of South Australia and Victoria described in Chapter 5.

Despite these achievements at the state and territory level, the inevitable question that arises is to what extent they collectively function as a national topographic database at the present time. A partial answer to this question can be found in the measures taken by the consortium of public sector mapping agencies (PSMA) to establish a Commonwealth-wide digital database for the Australia Bureau of Statistics 1996 Census of Population and Housing. For this reason, the establishment of this consortium with a mandate to return economic benefits to the nation through the preparation and exploitation of a national topographic database must be regarded as a significant move towards the creation of a core database for Australia as a whole.

The provision of core data in the Netherlands essentially reflects the split in responsibilities for small-scale mapping which lies in the hands of the Topografische Dienst and large-scale mapping and cadastral responsibilities which is in the hands of the Cadastre. As a result, two separate core databases are under development to meet the needs of two different groups of users. The GBKN (Grootschalige Basiskaart Nederland) project began in 1975 with an agreement between the Cadastre and some of the main users of large-scale map data in the Netherlands such as the utilities and local government to develop a large-scale map series mainly at 1:1,000 scale. As a result of a further agreement in 1992 between these parties regarding the financing of the project, considerable progress has been made in the conversion of data to digital format and this may be completed by the end of 1997. A distinctive feature of this project is that it involves a number of separate databases prepared by provincial private public partnerships to varying specifications. Consequently there may be as many as 75 subsets of the GBKN when it is completed.

The other core database is the 1:10,000 digital topographic database that is being created by the Topografische Dienst. At the beginning of 1996, some 60 percent of the Netherlands was covered by the digital TOP10vector structure and it is envisaged that all 675 sheets will be available in digital format by the end of 1997. Although this database is designed primarily to meet the needs of some of the big users in central government rather than utilities and local government users as was the case with the GBKN, the need to harmonise these two data sets must be regarded as a matter of some urgency.

Given the devolution of land titles registration and mapping responsibilities to state and local governments in the United States and the absence of common procedures or codes of practice, there is little likelihood that a comprehensive national digital topographic database that fulfils the requirements set out in the Framework document will come into being in the foreseeable future. Nevertheless, considerable progress has been made in some areas, most notably with respect to the transportation and hydrography layers for the 1:100,000 scale maps which are already available in digital form as a result of the agreement reached between the Bureau of the Census and the National Mapping Division of the US Geological Survey regarding the 1990 census. Similar arrangements are also being made between these and other parties for the 2000 census.

Despite these difficulties, the Federal Geographic Data Committee is fortunate in being able to draw upon the experience of many state-wide coordination efforts over the last few years and take advantage of the lessons to be learnt from a wide range of information-sharing partnerships. Consequently the main emphasis of FGDC activities with respect to core data has shifted to facilitating different kinds of information-sharing partnership at both the federal and the state and local government levels.

In summary, then, the findings of this analysis show that there are considerable differences between the four case-study countries with respect to the core data sets that have been created as a key element of national geographic information strategy. These closely reflect the distribution of responsibilities between the levels of government summarised in Table 7.2. Britain has had all seven core features defined by the US Framework document in place since May 1995 with the exception of cadastral information. Most of the Australian state and territories already have core databases in place which meet all seven features and the first steps have been taken towards the creation of an operational national spatial database at the Commonwealth level. The Netherlands is currently engaged in creating two separate core databases to meet the needs of small- and large-scale map users respectively and there are as yet no procedures in hand for harmonising them. Because of the devolution of responsibilities to state and local government, the United States faces a mammoth task and it is unlikely that a comprehensive core database will come into being in the foreseeable future. Nevertheless, some progress has been made in creating core databases with respect to hydrography and transportation.

Metadata

By comparison with the other two key elements of national geographic information strategy, the development of national (and international) metadata services is a task that is technically challenging but less demanding in terms of its needs for political commitment and human and other resources. There is general agreement in all four case-study countries that such services are an important element of national geographic information strategy, but only two of them (Britain and the United States) had services in operation in 1996. Nevertheless, work is under way in both the Netherlands and Australia which should lead to operational services in the near future.

The Ravi launched its National Clearing House for Geographic Information project in May 1995 and has formulated a strategy which builds on the experience of earlier metadata initiatives undertaken by various organisations in the Netherlands. The main objective of this project is to develop a uniform entry point to geographic information in the Netherlands. Similarly in Australia one of the components of the current ANZLIC strategic plan is to develop a national land and geographic data directory system for Australia and New Zealand.

The two operational systems present two contrasting models of national metadata services. Britain's Spatial Information Network Enquiry Service (SINES) is a simple metadata service which has been administered by Ordnance Survey since 1994. It is a database that contains details of almost 600 spatially referenced data sets held by more than 40 government departments and related bodies. SINES can be accessed by telephone, fax, email or directly through the World Wide Web. To find out what data services exist on a particular topic or for a particular area, users can search the database using key words. In its current form SINES gives a useful overview of what spatial information is held by government departments and also gives contact points for each data set so that potential users can obtain further information about their availability if they need.

In essence, then, the concept of SINES is similar to that of the Yellow Pages telephone directory. It is a single database that users can search to obtain information about the information or services that they need. It is then up to the user to take the necessary steps to contact the data (or service) provider to find out more about what is available and how much it costs.

The National Geospatial Data Clearinghouse which is being developed as an integral part of America's National Spatial Data Infrastructure differs fundamentally from metadata services such as SINES in that it is not a single database but a network of interlinked databases maintained by different custodians that can be accessed over the Internet. In mid-1996 it had 11 fully active nodes in operation and a similar number under development. As the example of the North Carolina node shows, each of these nodes in itself may involve several different participants from both the public and the private sectors. The key to the operation of a distributed network of databases such as this is the metadata standards that have been developed by the FGDC to provide a common format for describing the quality and characteristics of geospatial data throughout the Clearinghouse. Since 1995 all federal agencies have been required to use this standard to document their new data.

It should also be noted that, given the marginal cost/copyright free dissemination strategies adopted by many federal agencies in the United States, it is also increasingly possible for users to access the basic data sets themselves over the Internet within metadata services such as those provided by the National Geospatial Data Clearinghouse. For example, the National Mapping Division of the US Geological Survey has already made some of its products freely available over the Internet. These include the 1:100,000 scale transportation and hydrography layers developed for the 1990 census.

SOME LESSONS FOR OTHER COUNTRIES

The findings of the comparative evaluation show that there are considerable differences between the experiences of the four case-study countries with respect to the development of national geographic information strategies (or national spatial data infrastructures). These largely reflect the distribution of responsibilities between the different levels of government and the particular institutional context of each country as well as its historical and geographical setting. Given these circumstances it is clearly apparent that no single 'best' model of a national geographic information strategy emerges from this analysis which could be implemented without considerable modifications by other countries. Nevertheless a number of useful lessons can be learnt from the experiences of the four case-study countries which should be taken into account by other governments in the development of their national geographic information strategies. These can be summarised as follows with respect to the three key elements of such strategies.

Coordination

Although it has been to some extent taken for granted in much of the preceding analysis, a critical factor in the development of any national geographic information strategy is an awareness, not only within government but within the public at large, of the potential opportunities that have been opened up by recent developments in geographic information technology. Closely linked to this awareness is the need to recognise that geographic information is a national asset which needs to be effectively coordinated and managed to protect the national interest. Without such awareness, particularly within government, it is unlikely that much progress will be made towards developing a national geographic information strategy.

The second main lesson that can be drawn from the analysis is that political commitment is an important component of any national geographic information strategy. While it is unlikely that many countries will be able to muster the impressive level of political support that has been given to the development of the National Spatial Data Infrastructure in the United States, some measure of agreement within government is prerequisite for national-level coordination. The experience of the case-study countries shows that there are a number of alternative models for a coordinating body. In the United States geographic information coordination is handled through a high-level federal interagency committee. In Australia it is dealt with through an agency set up primarily to coordinate the transfer of geographic information between the different levels of government. In the Netherlands (and also in Britain if the proposals contained in the Chorley Report had been accepted by government) coordination is in the hands of an independent body with strong links to government.

Even with strong political support, however, the development and implementation of a coordinated national geographic information strategy is likely to depend heavily on the extent to which there is a spirit of collaboration between the various government agencies involved with geographic information. This has been essential for the successful operation of the pragmatic approach that has been adopted by the Ravi in the Netherlands and has also been an important factor in the work of the Interdepartmental Group on Geographic Information in Britain. This is the third lesson that can be learnt from the experience of the four case-study countries.

The fourth lesson is that, given a high level of awareness of the importance of geographic information for the national interest, together with the political support for coordination and a spirit of collaboration among (most of) those government agencies involved, the actual task of coordination itself is not expensive relative to the overall national investment in geographic information by governments. For example, the Office of Management and Budget in the United States estimates that federal agencies alone spent $4 billion annually to collect and manage domestic geospatial data (FGDC, 1994a, p. 2). This sum is of a very different order to that required to maintain the 12 staff positions to support the Federal Geographic Data Committee.

Core Data

One of the most important lessons that can be learnt from the above discussion is the extent to which the creation of core databases within a national geographic information strategy is linked to the way in which the main responsibilities for geographic information are distributed between different levels of government. Britain and the Netherlands (and, to a lesser extent, Australia) have clear advantages in this respect because of the extent to which these responsibilities are centralised over the United States which faces a mammoth task as a result of its devolved system of responsibilities. However, it should be noted that these advantages are not enough in themselves without the necessary vision in the agencies involved. There are plenty of examples of countries throughout the world that share some or all of these advantages without sharing the same vision with respect to the strategic importance of core data.

In all four countries, however, there is a need to build up partnerships for the creation and maintenance of core databases in the national interest. For this reason a great deal of effort is being devoted within the American National Spatial Data Infrastructure programme to promoting partnerships of all kinds to create core data sets between the agencies involved. The Netherlands offers an interesting partnership model in its GBKN large-scale mapping project. This is based on a formal agreement between two central agencies, the Cadastre and Dutch Telecom, the national associations of energy and water supply authorities and the association of local authorities, to establish the necessary public private partnerships needed to complete the national large-scale digital map coverage for Netherlands. As a result of regional and local variations in these partnerships it is likely that there will be over 75 separate subsets of this coverage by the time it is finished. Britain is also experimenting with partnership models in connection with the National Land Information Service and the National Geospatial Data Framework, while Australia's consortium of public sector mapping agencies has taken the first steps towards the integration of the digital topographic and cadastral databases held by the states and territories.

It will be apparent that partnerships such as these can only be effective from the national point of view if the databases are created using common data standards for data formatting and exchange. Common standards are also the key to Ordnance Survey's vision of how its national topographic database might be linked to other spatial data held by government departments in the National Geospatial Data Framework. Consequently, matters relating to standards feature prominently on the agendas of both international as well as national agencies at the present time through the activities of CEN TC 287 in Europe (European Committee of Standards technical committee for geographic information) and the work of the International Standards Office TC 211 for geographic information.

Metadata

It has been argued that the development of metadata services is technically challenging but less politically demanding than either of the two other key elements of a national geographic information strategy. Despite these conclusions, the importance of developing appropriate metadata services which enable users to find the information they want should not be underestimated.

Like core data, the key to the development of such services, whether or not they follow the Yellow Pages model of Ordnance Survey's SINES service, or the distributed model that is being implemented as part of the US National Geospatial Data Clearinghouse, is the need for common standards of documentation. There is also a political and an organisational dimension to metadata which is evident in the US government's decision in 1995 to require all federal agencies to use the FGDC metadata standard to document their new data.

From the technical standpoint the situation is changing rapidly at the present time as a result of the growing use of the Internet and the potential that this opens up for the transfer of the data themselves as well as information about the data to potential users. This in turn raises important issues with respect to matters such as access and charging which have yet to be resolved with respect to the electronic transfer of geographic information. Consequently it will be particularly necessary to monitor future developments closely in the case study countries and elsewhere in the world to keep track of what is happening in this rapidly changing component of national geographic information strategy.

References

ABEL, D.J. (1996) Australian experience in establishing a geospatial data infrastructure, in *Proceedings of the National Geospatial Database Seminar*, Southampton: Ordnance Survey.

AGI (1993) *Report by the Copyright Working Party*, London: Association for Geographic Information.

(1995) Opening up public data, *AGI Newsletter*, **5** (3), 1–2.

AKKERS, B., OVERDUIN, T., OVERMARS, B. & VOET, P.V.D. (1995) GIS ondersteuning bij hoogwater in Gelderland, *Vi Matrix*, **3** (4), 8–11.

ALEXANDER, D.P. (1995) Geographic information planning: targeting the benefits, *Proc. AURISA '95*, Canberra: Australasian Urban and Regional Information Systems Association, pp. 504–14.

ALIC (1990a) Data custodianship/trusteeship, *Issues in Land Information Management Paper No. 1*, Canberra: Australia Government Printing Service.

(1990b) A general guide to copyright, royalties and data use arrangements, *Issues in Land Information Management Paper No. 2*, Canberra: Australia Government Printing Service.

(1990c) Charging for land information, *Issues in Land Information Management Paper No. 3*, Canberra: Australia Government Printing Service.

(1990d) Access to government land information – commercialisation or public benefit? *Issues in Land Information Management Paper No. 4*, Canberra: Australia Government Printing Service.

ANDERSON, K.E. (1996) A strategic perspective of geographic information systems in the United States. In *Proc. International Seminar on Strategies for National GIS Development*, Anyang: Korea Research Institute for Human Settlements, pp. 1–10.

ANZLIC (1992a) *Land Information Management in Australasia 1990–1992*, Canberra: Australia Government Publishing Service.

(1992b) Privacy, confidentiality and access to information, *Land Information Management Paper No. 5*, Canberra: Australia Government Printing Service.

(1994) *Strategic Plan 1994–1997*, Canberra: Australia Government Publishing Service.

(1996) *ANZLIC Guidelines: Core Metadata Elements Version 1*, Canberra: Australia New Zealand Land Information Council.

VAN ASPEREN, P. (1996) 'Digital updates at the Topografische Dienst', paper presented at The XVIIIth ISPRS Congress, Vienna.

BABBITT, B. (1994) GIS World interview, *GIS World*, **7** (9), 31–3.

(1996) 'Establishing roots in our landscapes of complexity', keynote address to ESRI – Arc/Info Users Conference, Palm Springs, CA.

BAKER, G. (1995a) 'Towards a national land and information infrastructure', paper presented at Australian Mapping Circle Conference, Armidale.

(1995b) 'The role of public interest policy and co-ordination in AUSLIG', unpublished paper.

BARNES, A.K. (1995) 'A socio and economic database (SEDB) for South Australia', unpublished Masters thesis, Flinders University of South Australia.

BESEMER, J. (1994) 'The Cadastre as an independent public body', presentation at the opening ceremony of the ITC–TUD Centre for Cadastral Studies.

BLAKEMORE, M.J. (1991) Managing an operational GIS: the UK National Online Manpower Information System (NOMIS), in Maguire, D.J., Goodchild, M.F. and Rhind, D.W. (Eds) *Geographical Information Systems: Principles and Applications*, pp. 503–13, Harlow: Longman Scientific and Technical.

BLAKEMORE, M. and SINGH, G. (1992) *Cost Recovery Charging for Geographic Information: A False Economy*? London: Gurmukh Singh & Associates.

BRANSCOMB, A.W. (1994) *Who Owns Information? From Privacy to Public Access*, New York: Basic Books.

(1995) Public and private domains of information: defining the legal boundaries, *Bulletin of the American Society for Information Science*, Dec./Jan., 14–18.

BRYAN, N.S. (1993) A review of pricing and distribution strategies: local government case studies, in Bamberger, W.J. and Bryan, N.S. (Eds) *Marketing Government Information: Issues and Guidelines*, pp. 75–84, Washington DC: Urban and Regional Information Systems Association.

BUREAU OF THE CENSUS (1996) *The plan for Census 2000*, Washington DC: US Department of Commerce.

BURNHILL, P. (1991) Metadata and cataloguing standards: an eye on the spatial, in Medycki Scott, D., Newman, I., Ruggles, C. and Walker, D. (Eds) *Metadata in the Geosciences*, Loughborough: Group D Publications.

CEC (1990) *Access to Environmental Information*, Directive 90/313/EEC, Brussels: Commission of the European Communities.

(1996) *Legal Protection of Data Bases*, Directive 96/9/EEC, Brussels: Commission of the European Communities.

CHAN, T.O. & WILLIAMSON, I. (1995) Justification of GIS as an infrastructure investment: some observations regarding GIS management in Victoria, *Proc. AURISA 95*, Canberra: Australasian Urban and Regional Information Systems Association, pp. 492–503.

CHANCELLOR OF THE DUCHY OF LANCASTER (1996) *Next Steps Agencies in Government Review 1995*, Cm 3164, London: HMSO.

CLEVELAND, H. (1982) Information as a resource, *The Futurist*, **16** (Dec.), 34–9.

(1985) The twilight of hierarchy: speculations on the global information society, *Public Administration Review*, **7**, 1–31.

COMMITTEE OF PUBLIC ACCOUNTS (1994) *Data Protection Controls and Safeguards*, Twenty-ninth Report of the Committee of Public Accounts, London: HMSO.

COOMBES, M. (1995) Dealing with census geography: principles, practices and possibilities, in Openshaw, S. (Ed.) *Census User's Handbook*, Cambridge: Geoinformation International, pp. 111–32.

COOPERS AND LYBRAND (1996) *Economic Aspects of the Collection, Dissemination and Integration of Government's Spatial Information*, Southampton: Ordnance Survey.

CRONIN, B. (1984) Information accounting, in van der Laan, A. & Winters, A.A. (Eds) *The Use of Information in a Changing World*, Amsterdam: Elsevier.

CSDC (1994) *Commonwealth Spatial Data Committee Annual Report 1993–94*, Canberra: Australia Government Printing Service.

CUSHNIE, J. (1994) A British standard is published, *Mapping Awareness*, **8** (5), 40–3.

DALE, P. (1994) Towards a national land information system, in Green, D.R., Rix, D. & Cadoux Hudson, J. (Eds) *Geographic Information 1994*, pp. 233–6, London: Taylor & Francis.

DANDO, L.P. (1993) A survey of open records laws in relation to recovery of database development costs: an end in search of a means, in Bamberger, W.J. & Bryan, N.S. (Eds) *Marketing*

Government Information: Issues and Guidelines, pp. 5–22, Washington DC: Urban and Regional Information Systems Association.

Dansby, H.B. (1992) Survey and analysis of state GIS legislation, *GIS Law*, **1** (1), 7–13.

Data Protection Registrar (1993) *Ninth Report of the Data Protection Registrar*, London: HMSO.

Davey, A. & Murray, K. (1996) 'Update on the National Geospatial Database: collaboration between organisations', paper presented at the AGI Conference at GIS 96, Birmingham, 24–6 September.

Department of Administrative Services (1996) DAS Budget 1996–97, http://www.das.gov.au/~corpcomm.world/media/budget/

Department of the Environment (1987) *Handling Geographic Information: Report of the Committee of Enquiry chaired by Lord Chorley*, London: HMSO.

(1992) *Environmental Information Regulations*, London: HMSO.

Department of Lands (1986) *The Measure of the Land*, Adelaide, South Australia: Department of Lands.

DGXIII (1996) *GI2000: Towards a European Policy Framework for Geographic Information*, Luxembourg: Commission of the European Communities DGXIIIE.

DTI (1990) *Government-held Tradeable Information*, London: Department of Trade and Industry.

Eaton, J.J. & Bawden, D. (1991) What kind of resource is information? *International Journal of Information Management* **11**, 156–65.

van Eechoud, M.M. (1995) *Copyright on Geographic Information in the Netherlands*, Amersfoort: Netherlands Council for Geographic Information.

(1996) *Legal Protection of Geographic Information*, Amersfoort: European Umbrella Organisation for Geographic Information.

Epstein, E.F. & McLaughlin, J.D. (1990) A discussion of public information: Who owns it? Who uses it? Should we limit access? *ACSM Bulletin*, Oct., 33–8.

Executive Office of the President (1994) Coordinating geographic data acquisition and access: the National Spatial Data Infrastructure. Executive Order 12906, *Federal Register 59*, 17671–4.

FGDC (1994a) *The 1994 Plan for the National Spatial Data Infrastructure: Building the Foundations of an Information based Society*, Reston, VA: Federal Geographic Data Committee, USGS.

(1994b) *The National Geospatial Data Clearing House: A Report on Federal Agency Activities within the First Six Months of Executive Order 12906*, Reston, VA: Federal Geographic Data Committee, USGS.

(1995) *Development of a National Digital Geospatial Framework*, Reston, VA: Federal Geographic Data Committee, USGS.

Flaherty, D.H. (1989) *Protecting Privacy in Surveillance Societies*, Chapel Hill, NC: University of North Carolina Press.

Frederick, D. (1995) Coordination of surveying, mapping and related spatial data activities, in Onsrud, H.J. & Rushton, G. (Eds) *Sharing Geographic Information*, pp. 355–62, New Brunswick, NJ: Centre for Urban Policy Research.

Garnsworthy, J. & Hadley, C. (1994) SINES – pointing you in the right direction, *IGGI News*, **1**, 3.

GBKN (1992) *Raamovereenkomst Samenwerkingsverband GBKN*, Amersfoort: GBKN.

Gilbert, E.W. (1958) Pioneer maps of health and disease in England, *Geographical Journal*, **124**, 172–83.

Goddard, J. (1989) Editorial preface, in Hepworth, M., *Geography of the Information Economy*, London: Belhaven.

Gore, A. (1993) *From Red Tape to Results: Creating a Government that Works Better and Costs Less, Report of the National Performance Review*, Washington DC: US Government Printing Office.

GSS (1995) *Charging Guidelines for Statistical Products and Services*, London: Government Statistical Service.

HILMER, F.G. (1993) *National Competition Policy: Report by the Independent Committee of Inquiry*, Canberra: Australia Government Printing Office.

HM LAND REGISTRY (1996) *HM Land Registry Executive Agency: Annual Report and Accounts 1995–96*, London: HMSO.

HM TREASURY (1989) *The Financing and Accountability of Next Steps Agencies*, Cm 914, London: HMSO.

HOOKHAM, C. (1995) By the people, for the people – availability and pricing of government data, *Mapping Awareness*, **9** (2), 20–4.

HOUSE OF LORDS SELECT COMMITTEE ON SCIENCE AND TECHNOLOGY (1983) *Remote Sensing and Digital Mapping*, London: HMSO.

HUXHOLD, W.E. (1991) *An Introduction to Urban Geographic Information Systems*, Oxford: Oxford University Press.

JOHNSON, A. (1995) Spatial information industry development within South Australia, *Proc. AURISA '95*, Canberra: Australasian Urban and Regional Information Systems Association, pp. 16–25.

KADASTER (1996) *Kadaster Jaarverslag 1995*, Apeldoorn: Kadaster.

KARJALA, D.S. (1995) Copyright in electronic maps, *Jurimetrics Journal*, **35**, 395–415.

LARNER, A. (1992) Encouraging data market growth, *Mapping Awareness*, **6** (9), 29–31.

LAUGHLIN, G.P. (1994) Spatial statistics and the future: an ABS perspective, *Proc. AURISA '94*, Canberra: Australasian Urban and Regional Systems Association.

LENGKEEK, W. (1996) 'The adoption of a running statistical process to a harmonised geoinformation data source (TOP10vector)', paper presented at the work session on GIS, Washington DC.

LESLIE, S. (1994) The Association for Geographic Information, *IGGI News*, **1**, 6.

MAFFINI, G. (1990) The role of public domain databases in the growth and development of GIS, *Mapping Awareness*, **4** (1), 49–54.

MANTHORPE, J. (1995) The Land Register and the NLIS, in Shand, P.J. & Ireland, P. (Eds) *The 1995 European GIS yearbook*, pp. 129–31, Oxford: NCC Blackwell and Blenheim.

MARX, R.W. (1990) Introduction to special issue on the Census Bureau's TIGER system, *Cartography and Geographic Information Systems*, **17**, 17–19.

MASON, R.O., MASON, F.M. & CULNAN, M.J. (1994) *Ethics of Information Management*, London: Sage.

MASSER, I. (1988) The development of geographic information systems in Britain: the Chorley report in perspective, *Environment and Planning*, **B 15**, 489–94.

MASSER, I., CAMPBELL, H.J. & CRAGLIA, M. (1996) (Eds) *GIS Diffusion: The Adoption and Use of Geographic Information Systems in Local Government in Europe*, London: Taylor & Francis.

MASSINGHAM, R. (1996) UK Standard Geographic Base update, *IGGI News*, **5**, 1–2.

MCLENNAN, W. (1995) Address to ACT Statistical Society, Knibbs Lecture, Canberra.

MCMASTER, P. (1991) Ordnance Survey: 200 years of mapping and on, *Mapping Awareness*, **5** (7), 9–13.

MINISTRY OF CONSTRUCTION AND TECHNOLOGY (1995) *A Master Plan for a National Geographic Information System (NGIS) in Korea*, Seoul: Ministry of Construction and Technology.

MOBBS, J. (1996) The PSMA and national client requirements in the Report of the PSMA/ICSM workshop on digital cadastral databases, Melbourne, 5–7 August, pp. 4–7.

VAN DER MOLEN, P. (1994) 'LIS/Cadastre and privacy', paper presented at XXth FIG Congress, Melbourne.

(1996) Monopoly in a customer driven way, Interview with *GIM, Geodetic Information Management*, **10** (3), 34–7.

MOM, P. (1996) Zelf registrerende GBKN gemeente geaccepteered, *Vi Matrix*, **4** (2), 14–17.

MOONEY, D.J. & GRANT, D.M. (1995) 'National spatial data infrastructure', paper presented at the Cambridge Conference of National Mapping Organisations, Cambridge, July.

NANSEN, B., SMITH, N. & DAVEY, A. (1996) A British National Geospatial Database: Part I – What it is and why we need it, and Part II – How it might be achieved, *Mapping Awareness*, **10** (3), 18–20 and **10** (4), 38–40.

NATIONAL RESEARCH COUNCIL (1990) *Spatial Data Needs: The Future of the National Mapping Program*, Mapping Sciences Committee, Washington DC: National Academy Press.

(1993) *Toward a Coordinated Spatial Data Infrastructure for the Nation*, Mapping Sciences Committee, Washington DC: National Academy Press.

(1994) *Promoting the National Spatial Data Infrastructure through Partnerships*, Washington DC: National Academy Press.

NORTH CAROLINA GEOGRAPHIC INFORMATION COORDINATING COUNCIL (1994) Strategic plan for geographic information in North Carolina, Raleigh, NC: North Carolina Geographic Information Coordinating Council.

OGDC (1995) The Office of Geographic Data Co-ordination, Information Paper 1/10/100/1, Melbourne: Office of Geographic Data Co-ordination.

OLIVER, A. (1996a) Group dynamics: IGGI forges ahead on government geodata, *Mapping Awareness*, **10** (3), 26–8.

(1996b) AGI/Government Roundtables: the outcome, *Proc. AGI 96*, London: Association for Geographic Information, pp. 1.3.1–1.3.3.

OMB (1990) *Coordination of Surveying, Mapping and Related Spatial Data Activities*, Circular A-16 revised, Office of Management and Budget, Washington DC: Executive Office of the President.

(1996) *Management of Federal Information Resources*, Circular A-130 revised, Office of Management and Budget, Washington DC: Executive Office of the President.

ONSRUD, H.J. (1992) In support of open access for publicly held geographic information, *GIS Law*, **1**, 3–6.

(1995) The role of law in impeding and facilitating the sharing of geographic information, in Onsrud, H.J. & Rushton, G. (Eds) *Sharing Geographic Information*, Brunswick NJ: Centre for Urban Policy Research, Rutgers University.

ONSRUD, H.J., JOHNSON, J.P. & LOPEZ, X. (1994) Protecting personal privacy in using geographic information systems, *Photogrammetic Engineering and Remote Sensing*, **60**, 1083–95.

ONSRUD, H.J., JOHNSON, J.P. & WINNECKI, J. (1996) GIS dissemination policy: two surveys and a suggested approach, *URISA Journal*, **8** (2), 8–23.

ONSRUD, H.J. & REIS, R.I. (1995) Law and information policy for spatial databases: a research agenda, *Jurimetrics Journal*, **35**, 377–93.

ONSRUD, H.J. & RUSHTON, G. (Eds) 1995. *Sharing Geographic Information*, Brunswick NJ: Centre for Urban Policy Research, Rutgers University.

OPCS (1994) *OPCS Business Plan 1994–95*, London: Office of Population Censuses and Surveys.

(1995) OPCS/CSO Merger, *Census Newsletter*, **34**, 2–4.

OPENSHAW, S. & GODDARD, J.B. (1987) Some implications of the commodification of information and the emerging information economy for applied geographical analysis in the UK, *Environment and Planning*, **A 19**, 1423–9.

OPIE, R. (1995) The Dutch Topographic Survey: from ministry to market place, *GIS Europe*, **4** (3), xvii–xix.

ORDNANCE SURVEY (1995) *Executive Agency Framework Document*, Southampton: Ordnance Survey.

(1996a) *Annual Report and Accounts 1995–96*, London: HMSO.

(1996b) *Mapping the Way to the New Millennium: Ordnance Survey Strategic Plan 1996–2001*, Southampton: Ordnance Survey.

(1996c) *Copyright 3: Digital Map Data*, Southampton: Ordnance Survey.

(1996d) *Results of the Consultation Exercise on 'The National Interest in Mapping'*, Southampton: Ordnance Survey.

OSBORNE, D. & GAEBLER, E. (1992) *Reinventing Government: How the Entrepreneurial Spirit is Transforming the Public Sector*, Reading, MA: Addison Wesley.

PERRITT, H.H. (1996) *Law and the Information Superhighway: Privacy, Access, Intellectual Property, Commerce, Liability*, Chichester: Wiley Law.

POLICY STUDIES INSTITUTE (1995) *Publaw III*, final report prepared by PSI London and CRID, Namur, London: Policy Studies Institute.

PRICE WATERHOUSE (1995) *Australian Land and Geographic Infrastructure Benefits Study*, Canberra: Australia Government Publishing Service.

RAVI (1992a) *Structuurschets vastgoedinformatie voorziening, deel 1: Bestuurlijke notitie*, Apeldoorn: Raad voor Vastgoedinformatie.

(1992b) *Structuurschets vastgoedinformatie voorziening, deel 2: Hoofdrapport*, Apeldoorn: Raad voor Vastgoedinformatie.

(1992c) *Structuurschets vastgoedinformatie voorziening, deel 3: Inventatisatie en analyse*, Apeldoorn: Raad voor Vastgoedinformatie.

(1994) *1:10,000 Kernbestand: haalbaarheidsonderzoek*, Amersfoort: National Council for Geographic Information.

(1995a) *Half-yearly Report on Ravi: Coordinating Body for Geographic Information in the Netherlands*, Amersfoort: Netherlands Council for Geographic Information.

(1995b) *Percel en persoon gerelateerd: GBA personidentificatie in de Kadastrale registratie*, Amersfoort: Netherlands Council for Geographic Information.

(1995c) *The National Geographic Information Infrastructure*, Amersfoort: Netherlands Council for Geographic Information.

REDFERN, P. (1987) *A Study of the Future of the Census of Population: Alternative Approaches*, Report CA-48–87–896-EN-C, Luxembourg: Office for Official Publications of the European Communities.

REPO, A.J. (1989) The value of information: approaches in economics, accounting and management science, *Journal of the American Society for Information Science*, **40**, 68–85.

RHIND, D. (1991) The role of the Ordnance Survey of Great Britain, *Cartographic Journal*, **28**, 188–99.

(1992) Data access, charging and copyright and their implications for geographic information systems, *International Journal of Geographic Information Systems*, **6**, 13–30.

(1995a) Spatial data from government, in Green, D.R. & Rix, D. (Eds) *The 1995 AGI Source Book for Geographic Information Systems*, pp. 101–6, Chichester: Wiley.

(1995b) Ordnance Survey to the Millennium: the new policies and plans of the National Mapping Agency, *Proc. AGI 95*, London: Association for Geographic Information.

(1996a) Economic, legal and public policy issues influencing the creation, accessibility and use of GIS databases, *Transactions in GIS*, **1**, 3–12.

(1996b) From military maps to market forces, *GIS Europe*, **5** (5), 20–2.

(Ed.) (1997) *The Framework for the World*, Cambridge: Geoinformation International.

RHIND, D.W. & MOUNSEY, H.M. (1989) The Chorley Committee, and 'Handling Geographic Information', *Environment and Planning*, **A 21**, 571–85.

SEDUNARY, M.E. (1987) Land information systems: the South Australian perspective, Masters thesis, Adelaide: South Australian Institute of Technology.

SERPELL, D. (1979) *Report of the Ordnance Survey Review Committee*, London: HMSO.

SMITH, B. (1996) The NLIS in 1996: the pilot project expands, *Mapping Awareness*, **10** (3), 22–4.

SMITH, B. & GOODWIN, R. (1996) National Land Information Service: feasibility study interim results, *Proc. AGI 96*, London: Association for Geographic Information, pp. 2.16.1–2.16.2.

SMITH, M. & THOMAS, E. (1996) 'National spatial data infrastructure: an Australian viewpoint', paper presented at the Emerging Global Spatial Data Infrastructure Conference, Bonn, Germany, 4–6 September.

SPERLING, J. (1995) Development and maintenance of the TIGER database: experiences in spatial data sharing at the US Bureau of the Census, in Onsrud, H.J. & Rushton, G. (Eds) *Sharing Geographic Information*, pp. 377–96, New Brunswick, NJ: Centre for Urban Policy Research.

STATISTICS NETHERLANDS (1995) *Statistics Netherlands in Brief*, Voorburg: Statistics Netherlands.

TOMMEL, D. (1995) Opening address by the secretary of state for housing, spatial planning and the environment, Joint European Conference and Exhibition on Geographical Information, The Hague.

TOSTA, N. (1995) Data policies and the National Spatial Data Infrastructure, in Onsrud, H.J. (Ed.) *Proc. Conference on Law and Information Policy for Spatial Data Bases*, pp. 106–113, Orono, ME: National Centre for Geographic Information and Analysis, University of Maine.

TOSTA, N. & DOMARATZ, M. (1996) The US National Spatial Data Infrastructure, in Craglia, M. & Couclelis, H. (Eds) *Geographic Information Research: Bridging the Atlantic*, pp. 19–27, London: Taylor & Francis.

US GEOLOGICAL SURVEY (1994) *US Geological Survey: Earth Science in the Public Service*, Washington DC: US Government Printing Office.

(1996) Home pages, http://www.usgs.gov

VISSER, H. & LENGKEEK, W. (1995) Bodemgebruik per 1 Januari 1993 in der provincies Groningen, Friesland and Drenthe, *Maandstatistiek Landbouw*, **42** (5), 43–9.

WAN, W.Y. & WILLIAMSON, I. (1995a) A review of the digital cadastral databases in Australia and New Zealand, *Australian Surveyor*, **40**, 41–52.

(1995b) The users view of digital cadastral databases in Australia, *Australian Surveyor*, **40**, 53–62.

WARNECKE, L. (1992) *State Geographic Information Activities Compendium*, Lexington, KY: Council for State Governments.

WEGENER, M. & MASSER, I. (1996) Brave new GIS worlds, in Masser, I., Campbell, H.J. & Craglia, M. (Eds) *GIS Diffusion: The Adoption and Use of Geographic Information Systems in Local Government in Europe*, pp. 9–21, London: Taylor & Francis.

WILLIAMSON, I. & ENEMARK, S. (1996) Understanding cadastral maps, *Australian Surveyor*, **41**, 38–52.

YAMAURA, A. (1996) 'National spatial data infrastructure: an Asian viewpoint', paper presented at the Global Spatial Data Infrastructure Conference, Bonn, Germany, 4–6 September.

Index